The Britannica Guide to
Inventions
That Changed the Modern World

TURNING POINTS IN HISTORY

THE BRITANNICA GUIDE TO
INVENTIONS
THAT CHANGED THE MODERN WORLD

EDITED BY ROBERT CURLEY, MANAGER, SCIENCE AND TECHNOLOGY

Britannica®
Educational Publishing

IN ASSOCIATION WITH

ROSEN
EDUCATIONAL SERVICES

Published in 2010 by Britannica Educational Publishing
(a trademark of Encyclopædia Britannica, Inc.)
in association with Rosen Educational Services, LLC
29 East 21st Street, New York, NY 10010.

Copyright © 2010 Encyclopædia Britannica, Inc. Britannica, Encyclopædia Britannica, and the Thistle logo are registered trademarks of Encyclopædia Britannica, Inc. All rights reserved.

Rosen Educational Services materials copyright © 2010 Rosen Educational Services, LLC. All rights reserved.

Distributed exclusively by Rosen Educational Services.
For a listing of additional Britannica Educational Publishing titles, call toll free (800) 237-9932.

First Edition

Britannica Educational Publishing
Michael I. Levy: Executive Editor
Marilyn L. Barton: Senior Coordinator, Production Control
Steven Bosco: Director, Editorial Technologies
Lisa S. Braucher: Senior Producer and Data Editor
Yvette Charboneau: Senior Copy Editor
Kathy Nakamura: Manager, Media Acquisition
Robert Curley: Manager, Science and Technology

Rosen Educational Services
Jeanne Nagle: Senior Editor
Nelson Sá: Art Director
Matthew Cauli: Designer
Introduction by Jeri Freedman

Library of Congress Cataloging-in-Publication Data

The Britannica guide to inventions that changed the modern world / edited by Robert Curley.—1st ed.
    p. cm.—(Turning points in history)
"In association with Britannica Educational Publishing, Rosen Educational Services."
ISBN 978-1-61530-020-4 (library binding)
 1. Inventions—History. 2. Technology—History. I. Curley, Robert, 1955– II. Title: Guide to inventions that changed the modern world.
T15.B6827 2010
609—dc22

2009037539

Manufactured in the United States of America

On the cover: Wind turbines, like these near Palm Springs, Calif., are just one of several inventions that have changed energy generation and consumption around the world. *David McNew/Getty Images*

CONTENTS

Introduction	12
CHAPTER 1: COMMUNICATION	21
Cuneiform	21
Origins in Sumaria	21
Akkadian Cuneiform	23
The Papyrus Scroll	25
In Ancient Egypt	25
In Ancient Greece	27
The Papyrus Plant	28
The Greek and Latin Alphabets	29
The Greek Alphabet	29
The Latin Alphabet	32
Parchment and Vellum	34
The Codex	36
Paper	37
The Printing Press	40
Origins in China	40
Reinvention in Europe	42
The Gutenberg Bible	44
Photography	46
Antecedents	47
Heliography	48
Daguerreotype	49
George Eastman	50
The Telegraph	51
Preelectric Telegraph Systems	51
The Morse System	52
The Telephone	54
Early Sound Transmitters	55
Gray and Bell: The Transmission of Speech	56
The AT&T Desk Phone	61
The Phonograph	62
Edison's Phonograph	63

The Graphophone and Gramophone	63
The Race to Market	64
Coin Slots and Home Phonographs	66
Motion Pictures	68
Sequential Photographs	68
Edison's Kinetograph and Kinetoscope	69
The *Cinématographe*	71
Radio	72
Hertz: Radio-Wave Experiments	72
Marconi's Wireless Telegraph	73
The Fleming Diode and De Forest Audion	74
Transmission of Speech	75
Television	76
Mechanical Systems	76
Electronic Systems	79
Philo Farnsworth	81
The Transistor	82
Motivation and Early Radar Research	83
Innovation at Bell Labs	84
The Communications Satellite	87
The Personal Computer	90
The Apple II	94
The Internet	95
Virtual Communities	96
CHAPTER 2: TRANSPORTATION	102
The Wheeled Chariot	102
Sailing Ships	104
From Rowed Vessels to Sails	104
Types of Sails	106
The Chinese Junk	108
The Compass	109

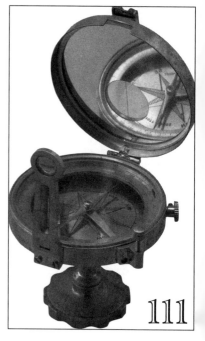

The Steamboat	113
Fulton's Steamboat	114
The Railroad	116
The Plateway	117
The Stockton and Darlington Railway	117
The Liverpool and Manchester Railway	119
The Rocket	120
The Bicycle	121
From Velocipedes to Boneshakers	121
The Penny-Farthing	122
The Safety Bicycle	123
The Automobile	124
The Four-Stroke Gasoline Engine	125
Karl Benz	126
Gottlieb Daimler	127
The Airplane	128
Early Experiments	129
The Wright Brothers	130
The Wright Flyer of 1903	131
The Space Launcher	133
Tsiolkovsky, Goddard, and Oberth	134
From the V-2 to the ICBM	135
The First "All-Civilian" System	137
The Jetliner	137
GPS	141
CHAPTER 3: POWER AND ENERGY	144
Controlled Fire	144
The Water Wheel	146
The Noria	148
The Windmill	149
The Steam Engine	153
Thomas Newcomen	156

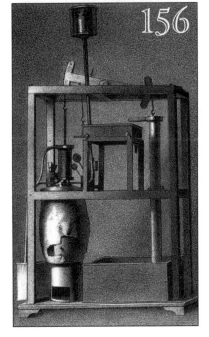

The Electric Battery	158
Galvani and Volta	158
Applications in Chemistry and Physics	160
The Electric Generator and Motor	161
The Incandescent Lightbulb	163
The Fluorescent Light	165
The Steam Turbine	167
Early Precursors	167
Modern Steam Turbines	168
The Gasoline Engine	170
The Jet Engine	173
Whittle, von Ohain, and the First Jet Aircraft	175
The Nuclear Reactor	178
The Laser	180
The Wind Turbine	184
The Solar Cell	187
The Fuel Cell	190
CHAPTER 4: BUILDING CONSTRUCTION AND CIVIL ENGINEERING	192
The Arch	192
Origins in the Post and Lintel	193
Development of the Arch	194
The Vault	195
Brick	196
Dams	198
The Romans	198
Early Dams of East Asia	199
The 15th to the 18th Century in Europe	200
The Aqueduct	201
The Roman Dome	204
The Pantheon	205

165

205

Plumbing	207
The Paved Road	209
Pierre-Marie-Jérôme Trésaguet	210
Thomas Telford	210
John Loudon McAdam	211
Reinforced Concrete	212
The Suspension Bridge	214
Dynamite	216
The Skyscraper	219
The Glass Curtain Wall	220
William Le Baron Jenney	224
The Elevator	225
Heating, Ventilation, and Air Conditioning	227

CHAPTER 5: MEDICAL MILESTONES 231

The Smallpox Vaccine	231
Edward Jenner and James Phipps	233
General Anesthesia	235
Pasteurization	237
X-ray Imaging	239
Ultrasound	240
Insulin	242
Antibiotics	244
Blood Transfusion	246
Charles Drew	248
Polio Vaccine	250
The Birth Control Pill	253
Heart Transplantation	256
The Artifical Heart	258
Genetic Engineering	262
Cloning	264
Early Cloning Experiments	265
Reproductive Cloning	266
A Sheep Named Dolly	268
Therapeutic Cloning	269

Chapter 6: Military Technology 272
The Spear 272
The Bow and Arrow 273
   The Early Bow 274
   The Crossbow 275
   Arrows 277
Gunpowder 278
Rifled Muzzle-Loaders 280
   Early Rifling 280
   Minié Rifles 281
The Submarine 284
   David Bushnell's Turtle 285
The Machine Gun 288
   Recoil 288
   Gas Operation 290
   Blowback 290
The Assault Rifle 291
The Tank 294
The Ballistic Missile 298
   The V-2 298
   The First ICBMs 300
   The Titan I and Titan II 301
Nuclear Weapons 303
   The First Atomic Bombs 303
   The First Hydrogen Bombs 310

Chapter 7: Observation and Measurement 314
The Gregorian Calendar 314
The Clock 317
The Watch 319
The Telescope 322
   The Hale Telescope 323
The Microscope 325
Radar 328
The Atomic Clock 331

CHAPTER 8: AGRICULTURE AND INDUSTRY	334
Bronze	334
Iron	337
The Plowshare	340
The Power Loom	344
The Basic Weaving Process	344
Automated and Power-Driven Looms	345
Canning	346
Refrigeration	348
Steel	350
The Bessemer Converter	352
Aluminum	354
Early Laboratory Extraction	354
Electrolytic Production	355
Sheet and Plate Glass	356
Rayon	359
Bakelite	361
Leo Baekeland	363
The Combine Harvester	364
Industrial Robots	366
Fullerenes	367
Glossary	371
For Further Reading	373
Index	375

*I*ntroduction

— INTRODUCTION —

The history of the human race can conceivably be called the story of innovation. Since time immemorial, inventions have served as a demarcation of civilization's advance. As humans have evolved, so, too, have the products of their ingenuity. From one of the first (fire) to several of the latest (the Internet, the fuel cell, genetic engineering), this book examines the inventions that have changed the world.

The creation of revolutionary products or practices has no timetable. Great minds give birth to innovations as inspiration hits, necessity requires, or happenstance would have it. However, there have been a number of specifically fruitful periods of invention throughout history. The Renaissance, for instance, was a period of great intellectual exploration, an era of innovation in the arts, philosophy, medicine, engineering, and science. Inventions of the Renaissance include the printing press and movable type. It was also at this time that Leonardo da Vinci (1452–1519) drew the first plans for a flying machine, a precursor to planes and jets that would be devised and developed centuries later.

The Industrial Revolution, which dates from the late 18th century through the early 19th century, was the great age of mechanization, characterized by the invention of machines for automating processes in manufacturing, mining, and agriculture. Automation made possible the creation of factories to mass-produce goods. Such goods could be sold at lower prices, which allowed ordinary people access to products they previously could not afford. Advances in transportation also allowed merchants to offer their goods and services over greater distances. Inventions within this timeframe include the power loom, the combine harvester, and the steam turbine, which powered the steamboats and train engines that delivered merchandise across countries and around the globe.

The Information Age, begun in the 20th century and bridging into the 21st, has changed the way people process information, conduct business, and communicate in general. It has had its share of important inventions as well. The computer, the cell phone, and the Internet have created the "Information Superhighway," where information on anything is available instantaneously, and it is possible to be in contact with anyone at any time, anywhere in the world. An argument can be made that the Information Age could be extended to include inventions such as the telephone, the telegraph, the radio, and television—all of which made the dissemination of information more efficient and effective.

This book details all of these inventions and more. What sets the included innovations apart and makes them noteworthy are the implications of their creation on cultures throughout the world. These are no flash-in-the-pan, fad items. Each profiled invention has had a potent and long-lasting impact on the way people live, work, and play—even their very survival.

Energy is required to create products, heat homes, power equipment, and fuel a vast array of human activities. The earliest form of energy that changed the world was fire—or, more accurately, the ability to control the use of fire. With this skill, people gained the power to cook food, harden formed clay into vessels, and illuminate the night, among other things. Evidence has been found for the possible use of fire nearly 1.5 million years ago.

Over the centuries, methods of harnessing other forms of energy were invented. Water was harnessed to power mills and factories via waterwheels; later, in the form of steam, it would be used to power industrial equipment and vehicles such as ships and trains. Wind was used to provide power as well, via old-fashioned windmills.

— INTRODUCTION —

In the 19th century, electricity changed the world almost as radically as the discovery of fire. Electricity provided people with the ability to have unlimited light at the touch of a switch, to cook food without the need for firewood or coal, and to power vast factories that sprang up everywhere during the Industrial Revolution. It contributed to the development of technologies such as the radio, television, refrigeration units, electric batteries and motors, and, of course, the lightbulb.

The search for new forms of energy continues, and the need for inventions that capture the power of nature proceeds apace. Today, scientists and ecologists are working to evolve new, cleaner, more Earth-friendly sources of energy using wind turbines, solar cells, and fuel cells, which use a chemical reaction to produce energy.

Like energy, food is a basic requirement for survival. For much of human history, food has been difficult to find or grow in sufficient quantity. Therefore, advances in agriculture that allow people to produce and harvest more food are critical. The development of automated equipment for harvesting crops in the nineteenth century made it possible to plant and harvest more crops in a shorter period of time. One example is the combine harvester. Agriculture is still undergoing major changes as a result of advances in genetic engineering, which has made possible the production of genetically modified crops that can grow in inhospitable areas, last longer, and resist insects.

Other inventions go beyond basic human needs to enhance human interaction. There are two basic types of communication. Passive communication presents content to listeners or viewers, but the communication flows only one way. The earliest, and possibly largest, leap in passive communication was the invention of writing. The earliest usage of the written word, dated circa 4000 to 3000 BCE, appears to have developed from a need to keep trade and accounting records. The first paperlike material to be developed was papyrus, made from the pulp of the papyrus plant and developed by the Egyptians around the 2nd millennium BCE.

Handwritten works were time-consuming and very expensive to produce. Therefore, written works could not be widely distributed until the invention of the printing press and movable type, which made it possible to make multiple copies of written documents. The printing press was invented by Johannes Gutenberg (c. 1398–1468) in Germany around 1440.

Active, or interactive, communication flows two ways, allowing people to exchange information. In the mid-1800s, the dual inventions of the telegraph and the telephone made it possible for people to instantly connect to others around town or across the country, sharing news and information, instead of relying on slower means of communication such as the mail.

However, the invention that has perhaps had the greatest effect on the field of active communication is the Internet, a global system of interconnected computer systems. Started in the mid-20th century by the U.S. Department of Defense as a means to maintain communication if a military attack destroyed conventional means of communication, it was later refined and made available to the general public.

The Internet has unquestionably changed the way that we conduct business, entertain ourselves, promote political objectives, and learn. It is now possible to find information on every topic imaginable, making the Internet an invaluable reference tool. It is also possible to buy every sort of product from online stores and auctions, creating the whole new field of e-commerce using online connectivity. The Internet allows people to view movies and episodes of TV shows, download music, and upload videos. Most importantly, people can communicate with others throughout the world.

Inventions in transportation have continuously increased the distances that people can travel and the safety with which they can make their voyages. Land transportation has evolved from horse-drawn carriages to gasoline-powered automobiles to fuel-efficient electric cars. Together, the inventions of the automobile, paved roads, and suspension bridges have allowed people to travel in hours distances that previously took days. Vehicles such as the bicycle, the steamboat, the railway system, and the jet plane have made travel more convenient for the common person, while space launchers have been ridden by a select few but have benefited humankind as a whole through the exploration of space.

The fields of construction and architecture also have their share of advancement. The arch, whose widespread use dates back to the ancient Romans, allowed building materials to span an open space. The Romans pioneered many other architectural advances, including aqueducts—raised conduits that carry water from distant sources into towns and cities—and paved roads that allowed early wheeled vehicles, such as the chariot, to travel easily.

One of the areas in which the effect of inventions is most immediately apparent is medicine. Life expectancy has increased dramatically from ancient times to the present. One of the main reasons for this is improved sanitation. Indoor plumbing accounts for a large part of the improved sanitation issue, but so does the ability to control disease-causing bacteria. A major discovery in the latter area was pasteurization, developed by the French chemist Louis

Pasteur (1822–1895). Pasteurization is a method of treating food with heat to kill bacteria, making it safe to eat.

The second way in which inventions have increased the quality and length of life involves the development of medicines. Antibiotics such as penicillin and vaccines such as the one for polio, which has saved vast numbers of youngsters from paralysis and death, are just a few of the pharmaceutical and medical advances that have allowed us to cure or control the symptoms of diseases that would have once been fatal.

Some inventions help human beings to better observe and understand how the world works; for example, the telescope. From the first telescopes, developed in the Netherlands in the 17th century, to the Hubble Space Telescope, placed in orbit in 1990, telescopes have given us the ability to observe and learn about the universe. Similarly, the microscope, which was also invented in the Netherlands at the end of the 16th century, has given us invaluable information about materials too small to see with the naked eye. The microscope has allowed us to learn how cells work and shown us the internal structure of the materials that make up our world. This knowledge has helped us create new treatments for disease and new types of materials for use in many industries.

Not all inventions save or somehow better lives. Most advances in military technology revolve around the development of weaponry, or defense against the weapons of adversaries. From the crossbow in the Middle Ages, which improved the distance and power of the conventional bow and arrow, to the 16th-century musket and the 20th-century atomic bomb, weapons have become ever more efficient and powerful. Military inventions have had some peaceful ramifications, however. Technology developed to bolster the delivery of missiles also led to the discovery of rocket boosters that lifted spacecraft, manned and unmanned, into orbit.

As we go through the next century, there is no doubt that new inventions in communication, transportation, energy, building, medicine, military technology, observation and measurement, and agriculture will continue to be developed. Some of these inventions will build on discoveries of the past. Others will be based on entirely new concepts. Regardless, there is little doubt that innovations of the future, like those covered in this book, will change the world in ways that we cannot begin to imagine.

Chapter 1: Communication

In today's interconnected world, communication—that is, the transmission of information for the purpose of creating understanding—can be seen as the product of a series of inventions and also as the basis of all further invention. Thanks to the invisible action of electrons working over the media of television, the telephone, and the Internet, our means of communication today have so extended our reach, broadened our vision, and expanded our knowledge that we are often said to live in the Age of Information. That age actually began thousands of years ago, with the first markings on clay tablets and sheets of papyrus.

CUNEIFORM

Cuneiform was the most widespread and historically significant writing system in the ancient Middle East, its active history comprising the last 3,000 years before the Common Era. The name *cuneiform*, a coinage from Latin and Middle French roots meaning "wedge-shaped," has been the modern designation for the writing system since the early 18th century.

Origins in Sumeria

The origins of cuneiform may be traced as far back as the end of the 4th millennium BCE. At that time the Sumerians, a people of unknown ethnic and linguistic affinities, inhabited southern Mesopotamia and the region west of the mouth of the Euphrates known as Chaldea. While it does not follow that they were the earliest inhabitants of the region or the true originators of their system of writing, it is to them that the first attested traces of cuneiform writing are conclusively assigned. The earliest written records in the Sumerian language are pictographic tablets from Uruk (Erech), evidently lists or ledgers of commodities identified by drawings of the objects and accompanied by numerals and personal names. Such word writing was able to express only the basic ideas of concrete objects. Numerical notions were easily rendered by the repetitional use of strokes or circles. However, the representation of proper names, for example, necessitated an early recourse to the rebus principle—i.e., the use of pictographic shapes to evoke in the reader's mind an underlying sound form rather than the basic notion of the drawn object. This brought about a transition from pure word writing to a partial phonetic script. Thus, for example, the picture of a hand came to stand not only for Sumerian šu ("hand") but also for the phonetic syllable šu in any required context. Sumerian words were largely monosyllabic, so the signs generally denoted syllables, and the resulting mixture is termed a word-syllabic script. The inventory of phonetic symbols henceforth enabled the Sumerians to denote grammatical elements by phonetic complements added to the word signs (logograms or ideograms). Because Sumerian had many identical sounding (homophonous) words, several logograms frequently yielded identical phonetic values and are distinguished in modern transliteration—(as, for example, ba, $bá$, $bà$, ba_4). Because a logogram often represented several related notions with different names (e.g., "sun," "day," "bright"), it was capable of assuming more than one phonetic value. This feature is called polyphony.

During the 3rd millennium BCE the writing became successively more cursive, and the pictographs developed into conventionalized linear drawings. Due to the prevalent use of clay tablets as writing material (stone, metal, or wood also were employed occasionally), the linear strokes acquired a wedge-shaped appearance by being pressed into the soft clay with the slanted edge of a stylus. Curving lines disappeared from writing, and the normal order of signs was fixed as running from left to right, without any word-divider. This change from earlier columns running downward entailed turning the signs on one side.

Akkadian Cuneiform

Before these developments had been completed, the Sumerian writing system was adopted by the Akkadians, Semitic invaders who established themselves in Mesopotamia about the middle of the 3rd millennium BCE. In adapting the script to their wholly different language, the Akkadians retained the Sumerian logograms and combinations of logograms for more complex notions but pronounced them as the corresponding Akkadian words. They also kept the phonetic values but extended them far beyond the original Sumerian inventory of simple types (open or closed syllables like *ba* or *ab*). Many more complex syllabic values of Sumerian logograms (of the type *kan, mul, bat*) were transferred to the phonetic level, and polyphony became an increasingly serious complication in Akkadian cuneiform (e.g., the original pictograph for "sun" may be read phonetically as *ud, tam, tú, par, laḫ, ḫiš*). The Akkadian readings of the logograms added new complicated values. Thus the sign for "land" or "mountain range" (originally a picture of three mountain tops) has the phonetic value *kur* on the basis of Sumerian but also *mat* and *šad* from Akkadian *mātu* ("land") and *šadû* ("mountain"). No effort was made until very late to alleviate the resulting confusion, and equivalent "graphies" like *ta-am* and *tam* continued to exist side by side throughout the long history of Akkadian cuneiform.

The earliest type of Semitic cuneiform in Mesopotamia is called the Old Akkadian, seen for example in the inscriptions of the ruler Sargon of Akkad (died c. 2279 BCE). Sumer, the southernmost part of the country, continued to be a loose agglomeration of independent city-states until it was united by Gudea of Lagash (died c. 2124 BCE) in a last brief manifestation of specifically Sumerian culture. The political hegemony then passed decisively to the Akkadians, and King Hammurabi of Babylon (died 1750 BCE) unified all of southern Mesopotamia. Babylonia thus became the great and influential centre of Mesopotamian culture. The Code of Hammurabi is written in Old Babylonian cuneiform, which developed throughout the shifting and less brilliant later eras of Babylonian history into Middle and New Babylonian types. Farther north in Mesopotamia the beginnings of Assur were humbler. Specifically Old Assyrian cuneiform is attested mostly in the records of Assyrian trading colonists in central Asia Minor (c. 1950 BCE; the so-called Cappadocian tablets) and Middle Assyrian in an extensive Law Code and other documents. The Neo-Assyrian period was the great era of Assyrian power, and the writing culminated in the extensive records from the library of Ashurbanipal at Nineveh (c. 650 BCE).

The expansion of cuneiform writing outside Mesopotamia began in the 3rd millennium, when the country of Elam in southwestern Iran was in contact with Mesopotamian culture and adopted the system of writing. The Elamite sideline of cuneiform continued far into the 1st millennium BCE, when it presumably provided the Indo-European Persians with the external model for creating a new simplified quasi-alphabetic cuneiform writing for the Old Persian language. The Hurrians in northern Mesopotamia and around the upper stretches of the Euphrates adopted Old Akkadian cuneiform around 2000 BCE and passed it on to the Indo-European Hittites, who had invaded central Asia Minor at about that time.

In the 2nd millennium the Akkadian of Babylonia, frequently in somewhat distorted and barbarous varieties, became a lingua franca of international intercourse in the entire Middle East, and cuneiform writing thus became a universal medium of written communication. The political correspondence of the era was conducted almost exclusively in that language and writing. Cuneiform was sometimes adapted, as in the consonantal script of the Canaanite city of Ugarit on the Syrian coast (*c.* 1400 BCE). Sometimes it was simply taken over, as in the inscriptions of the kingdom of Urartu or Haldi in the Armenian mountains from the 9th to 6th centuries BCE; the language is remotely related to Hurrian, and the script is a borrowed variety of Neo-Assyrian cuneiform. Even after the fall of the Assyrian and Babylonian kingdoms in the 7th and 6th centuries BCE, when Aramaic had become the general popular language, rather decadent varieties of Late Babylonian and Assyrian survived as written languages in cuneiform almost down to the time of Christ.

THE PAPYRUS SCROLL

The papyrus scroll of ancient Egypt is more nearly the direct ancestor of the modern book than is the clay tablet used in cuneiform writing. Papyrus as a writing material resembles paper. It was made from a reedy plant of the same name that flourishes in the Nile Valley.

IN ANCIENT EGYPT

Pliny the Elder, the Roman savant and author from the 1st century CE, gave an account of the manufacture of paper from papyrus. The fibrous layers within the stem of the plant were removed, and a number of these longitudinal strips were placed side by side and then crossed at right angles with another set of strips. The two layers formed a sheet, which was then dampened and pressed. Upon drying, the gluelike sap of the plant acted as an adhesive and cemented the layers together. The sheet was finally hammered and dried in the sun. The paper thus formed was pure white in colour and, if well-made, was free of spots, stains, or other defects. Although the sheets varied in size, ordinary ones measured about 5 to 6 inches (13 to 15 centimetres) wide. A number of these sheets were then joined together with paste to form a scroll, with usually not more than 20 sheets to a scroll. To make a book, the scribe copied a text on the side of the sheets where the strips of pith ran horizontally, and the finished product was rolled up with the text inside.

The use of papyrus affected the style of writing just as clay tablets had affected cuneiform. Scribes wrote on it with a reed pen or brush and inks of different colours. The result could be very decorative, especially when done in the monumental hieroglyphic style of writing, a style best adapted to stone inscriptions. The Egyptians created two cursive hands, the hieratic (priestly) and the demotic (a simplified form of hieratic suited to popular use), which were better adapted to papyrus.

Compared with tablets, papyrus is fragile, yet an example is extant from 2500 BCE; and stone inscriptions that are even older portray scribes with rolls. This amazing survival is partly the result of the dry climate of Egypt, in which some papyrus scrolls survived unprotected for centuries while buried in the desert sands. The practice of certain Egyptian funerary customs also contributed to the preservation of many Egyptian books. Obsessed by a concern with life after death, they wrote magical formulas on coffins and on the walls of tombs to guide the dead safely to the gates of the Egyptian underworld. When the space thus provided became insufficient, they entombed papyrus scrolls containing the texts. These mortuary texts are now described collectively as *The Book of the Dead*, although the Egyptians never standardized a uniform collection. Such books, when overlooked by grave robbers, survived in good condition in the tomb. Besides mortuary texts, Egyptian texts included scientific writings and a large number of myths, stories, and tales.

In Ancient Greece

The Greeks adopted the papyrus scroll and passed it on to the Romans. Although both Greeks and Romans used other writing materials (waxed wooden tablets, for example), the Greek and Roman words for "book" show identification with the Egyptian model. Greek *biblos* ("book") can be compared with *byblos* ("papyrus"), while the Latin *volumen* ("book") signified a scroll. It has been suggested that papyrus was continuously in use in Greece from the 6th century BCE, and evidence has been cited to indicate its use as early as 900 BCE. Objects called books are mentioned by ancient Greek writers as having been in use in the 5th century BCE. The oldest extant Greek rolls, however, date from the 4th century BCE.

The 30,000 extant Greek papyri permit a generalized description of the Greek book. Rolled up, it stood about nine or 10 inches high and was an inch or an inch and a half in diameter. When the book was unrolled it displayed a text written in the Greek alphabet in columns about three inches wide separated by inch-wide margins. In spite of the Greek proficiency in decorative arts, few surviving books are illustrated. Such illustrations as have survived were of the practical sort found in later scientific books.

Practicality was a mark of the Greek book. The alphabet, although not invented by the Greeks, was adapted and

THE PAPYRUS PLANT

The papyrus plant *(Cyperus papyrus)* was long cultivated in the Nile delta region of Egypt. The plant is formed into a smooth, thin writing surface that is also known as papyrus. It is a grasslike aquatic plant that has woody, bluntly triangular stems and grows up to about 15 feet (4.6 metres) high in quietly flowing water up to 3 feet (90 cm) deep. The triangular stem can grow to a width of as much as 2.5 inches (6 cm). Even now, the papyrus plant is often used as a pool ornamental in warm areas or in conservatories. The dwarf papyrus (*C. isocladus*, also given as *C. papyrus* "Nanus"), up to 2 feet (60 cm) tall, is sometimes potted and grown indoors.

Although writing material was the most important product of the papyrus plant, the ancient Egyptians also used the stem to make sails, cloth, mats, and cords. In fact, papyrus was cultivated and used for writing material by the Arabs of Egypt down to the time when the growing manufacture of paper from other plant fibres in the 8th and 9th centuries CE rendered papyrus unnecessary. By the 3rd century CE, papyrus had already begun to be replaced in Europe by the less-expensive vellum, or parchment, but the use of papyrus for books and documents persisted sporadically until about the 12th century.

Papyrus (Cyperus papyrus). Adrian Pingstone

stabilized by them as an instrument of verbal communication rather than of decorative purpose. Unlike the monumental Egyptian survivals in a decorative hand that sometimes exceeded 100 feet (30 metres) in length, Greek rolls seldom exceeded 35 feet (11 m) in length and featured little embellishment. Such a scroll was about as large as could be conveniently held in the hands to read, and it was big enough to contain a book of Thucydides or one of the longer New Testament Gospels. The average Greek book was shorter. Two books (here denoting a subdivision of a text) of Homer written in a later small hand fitted a 35-foot (11-m) scroll.

THE GREEK AND LATIN ALPHABETS

The invention of the alphabet is a major achievement of Western culture. It is also unique; the alphabet was invented only once, though it has been borrowed by many cultures. It is a model of analytic thinking, breaking down perceptible qualities like syllables into more basic constituents. The alphabet requires little of the reader beyond familiarity with its orthography (that is, the written symbols of the spoken sounds that make up a language). It allows the reader to decipher words newly encountered and permits the invention of spellings for new patterns of sound, including proper names (a problem that is formidable for nonalphabetic systems).

The word *alphabet*, from the first two letters of the Greek alphabet—*alpha* and *beta*—was first used, in its Latin form, *alphabetum*, by Tertullian (2nd–3rd century CE), a Latin ecclesiastical writer and church father, and by St. Jerome. The classical Greeks customarily used the plural of *to gramma* ("the letter"); the later form *alphabetos* was probably adopted under Latin influence.

THE GREEK ALPHABET

As in so many other things, the importance of the ancient Greeks in the history of the alphabet is paramount. All of the alphabets in use in European languages today are directly or indirectly related to the Greek. The Greek alphabet was of course the culmination of a long historical evolution; the main achievement of the Greeks was to provide representations for vowel sounds. Consonants plus vowels made a writing system that was both economical and unambiguous. The true alphabetic system has remained for 3,000 years, with only slight modifications, an unparalleled vehicle of expression and communication in and among the most diverse nationalities and languages.

The Greek alphabet derived from the North Semitic script in the 8th century BCE. The direction of writing in the oldest Greek inscriptions—as in the Semitic scripts—is from right to left, a style that was superseded by the *boustrophedon* (meaning, in Greek, "as the ox draws the plow"), in which lines run alternately from right to left and left to right. This change occurred approximately in the 6th century BCE. There are, however, some early Greek inscriptions written from left to right, and after 500 BCE Greek writing invariably proceeded from left to right.

The letters for *b*, *g*, *d*, *z*, *k*, *l*, *m*, *n*, *p*, *r*, and *t*—sounds common to the Semitic and Greek languages—were taken over without change. The principal Greek change arose in applying a script developed to represent a Semitic language, in which vowel sounds are of minor importance to the identity of a word, to a language in which such vowel differences are crucial to the identity of a word. In Greek, /kat/, /kit/, and /kot/ are entirely different words, while in Semitic languages they would be the same word in different grammatically inflected forms. The Greek addition of vowels to the alphabet to make it an analogue of the sound pattern produced a writing system that was both manageable and accurate.

The eastern and western subdivisions were the two principal branches of the early Greek alphabet. The Ionic alphabet was the most important of the eastern variety, which also included the Greek alphabets of Asia Minor and the adjacent islands, of the Cyclades and Attica, of Sicyon and Argos, and of Megara, Corinth, and the Ionian colonies of Magna Graecia. A secondary branch of the eastern subdivision was made up of the alphabets used on the Dorian islands of Thera, Melos, and Crete. The alphabets of Euboea (Chalcidian), Boeotia, Phocis, Locris, Thessaly, the Peloponnesus (except its northeastern part), and of the non-Ionian colonies of Magna Graecia belonged to the western subdivision.

The differences between all these local alphabets involved variations in detail rather than essential structure, and gradually the Greek local alphabets became more and more similar. In 403 BCE the Ionic alphabet of Miletus was officially adopted in Athens and later also in the other states. By the middle of the 4th century BCE almost all the local alphabets had been replaced by the Ionic, which became the common, classical Greek alphabet of 24 letters.

Countless inscriptions have been discovered all over the Hellenic and Hellenistic world and beyond. They include official decrees, annals, codes of law, lists of citizens, civic rolls, temple accounts, votive offerings, ostraca (fragments of pottery), sepulchral inscriptions, coins, lettering on vases, and so forth. These, along with many thousands of Greek manuscripts, both ancient and medieval, serve as sources for the studies known as Greek epigraphy and Greek paleography and are of untold importance for all branches of ancient history, philology, philosophy, and other disciplines.

The most direct offshoots from the Greek alphabet were those adapted to the languages of the non-Hellenic peoples of western Asia Minor in the 1st millennium BCE: the scripts of the Lycians, Phrygians, Pamphylians, Lydians, and Carians. More significant, however, were the European offshoots. In Italy, two alphabets derived directly from the Greek: the Messapian (Messapic) and the Etruscan.

The Latin Alphabet

The adaptation of the Etruscan alphabet to the Latin language probably took place some time in the 7th century BCE. From this century there is a gold brooch known as the Praeneste Fibula (preserved in the Museo Preistorico Etnografico Luigi Pigorini in Rome). The inscription, written in an early form of Latin, runs from right to left and clearly reads: "MANIOS MED FHEFHAKED NUMASIOI" (in Classical Latin: "*Manius me fecit Numerio*," meaning "Manius made me for Numerius"). Dated not much later than this is a vertical inscription on a famous cippus (small pillar) in the Roman Forum, and the Duenos inscription on a vase found near the Quirinal (a hill in Rome) probably dates to the 6th century BCE. It is also written from right to left.

The original Etruscan alphabet consisted of 26 letters, of which the Romans adopted only 21. They did not retain the three Greek aspirate letters (*theta, phi*, and *chi*) in the alphabet because there were no corresponding Latin sounds but did employ them to represent the numbers 100, 1,000, and 50. Of the three Etruscan *s* sounds, the Romans kept what had been the Greek *sigma*. The symbol that represented the aspirate later received the shape *H* as it did in Etruscan. *I* was the sign both of the vowel *i* and the consonant *j*. *X* was added later to represent the sound *x* and was placed at the end of the alphabet. At a later stage, after 250 BCE, the seventh letter, the Greek *zeta*, was dropped because Latin did not require it, and a new letter, *G*, made by adding a bar to the lower end of *C*, was placed in its position.

After the conquest of Greece in the 1st century BCE, a large number of Greek words were borrowed by the Latin language. At that time the symbols *Y* and *Z* were adopted from the contemporary Greek alphabet, but only to transliterate Greek words; hence, they do not appear in normal Latin inscriptions. They were placed at the end of the alphabet, and the Latin script thus became one of 23 symbols.

A few permanent additions or, rather, differentiations from existing letters occurred during the Middle Ages, when the signs for *u* and *v*, and *i* and *j*, previously written interchangeably for either the vowel or the consonant sound, became conventionalized as *u* and *i* for vowels and *v* and *j* for consonants. *W* was introduced by Norman scribes to represent the English sound *w* (a semivowel) and to differentiate it from the *v* sound.

The connection of the capital letters of modern writing with the ancient Semitic-Greek-Etruscan-Latin letters is evident even to a layman. Indeed, the Roman capital letters, a form of writing that was used under the empire with unparalleled effectiveness for monumental purposes, became a byword for precision and grandeur. The connection of the minuscules (i.e., the small letters) with the ancient Latin letters is not as evident, but in fact both the majuscules (i.e., the large, or capital, letters) and the minuscules descended from the same ancient Latin alphabet. The different shapes of the small letters are the result of a transformation of the ancient letters by the elimination of a part of the letter—as, for instance, *h* from *H* or *b* from *B*—or by lengthening a part of it—for instance, *d* from *D*. Moreover, the change of the Latin writing into the modern script was induced by the nature of the writing tool and the material of writing. It was the pen, with its preference for curves, that eliminated the angular forms; it was the papyrus, and still more the parchment or vellum (and, in modern times, paper) that made these curves possible.

PARCHMENT AND VELLUM

Parchment and vellum are materials prepared from the skins of animals. Strictly speaking, vellum is a finer quality of parchment prepared from calf skins, but the terms have been used interchangeably since the Middle Ages. The forerunner of parchment as a writing material was leather. Egyptian sources refer to documents written on leather as early as 2450 BCE, and a fragmentary Egyptian leather scroll of the 24th century BCE survives. However, leather was rarely used because papyrus was plentiful, and only a new, more thorough method of preparing parchment made possible the supplanting of the papyrus scroll. The skins of various animals—cattle, sheep, and goats being most common—were washed and divested of hair or wool. Then the skin was stretched tight on a frame, scraped thin to remove further traces of hair and flesh, whitened with chalk, and smoothed with pumice. Legend has it that parchment was invented as the result of a book-collecting rivalry between Ptolemy V of Egypt and Eumenes II of Pergamum (modern Bergama, Turkey) about 190 BCE. Fearing the library at Pergamum might outstrip the collections at Alexandria, Ptolemy placed an embargo on papyrus to prevent his rival from making any more books, whereupon Eumenes made parchment. The fact that both the Greek and Latin words for parchment mean "stuff from Pergamum" offers some support for the tradition.

Although parchment was used to produce book rolls, and although many early codices were made from papyrus, the new writing material facilitated the success of the codex. A sheet of parchment could be cut in a size larger than a sheet of papyrus; it was flexible and durable, and it could better receive writing on both sides. These qualities were important. In making a parchment or vellum codex, a large sheet was folded to form a folio of two leaves, a quaternion (quarto) of four, or even an octavo of eight. Gatherings were made from a number of these folded sheets, which were then stitched together to form a book. Because papyrus was more brittle and could not be made in large enough sheets, the folio collected in quires (i.e., loose sheets) was the limit of its usefulness. At the same time, because of the vertical alignment of the fibres on one side, papyrus was not well adapted for writing on both sides in a horizontal script.

For 400 years the scroll and the codex existed side by side. There are contemporary references to the codex book dating from the 1st century BCE; actual survivals date from the 2nd century CE, however. In the 4th century CE vellum or parchment as a material and the codex as a form became dominant, although there are later examples of rolls, and papyrus was occasionally used for official documents until the 10th century. The vellum of most early manuscripts, through the 6th century CE, was of good quality. After this, as demand increased, a great amount of inferior material came on the market, but by the 12th century, when large numbers of manuscripts were being produced in western Europe, a soft, pliant vellum was in vogue. In Constantinople, a sumptuous form was produced at an early date by dyeing the material a rich purple and lettering it in silver and gold, a practice condemned as a useless luxury in a well-known passage of St. Jerome. The purple dye was subsequently abandoned, but the practice of "illuminating" parchment manuscripts in gold, silver, and other tints flourished throughout the European Middle Ages.

THE CODEX

Codex Sinaiticus (British Museum, Add. MS. 43725, fol. 260). Courtesy of the trustees of the British Museum

The codex was the earliest type of manuscript to take the form of a modern book (i.e., a collection of written pages stitched together along one side). The codex form had several advantages over the papyrus scroll. It could be opened at once to any point in the text, enabled one to write on both sides of the leaf, and could contain long texts. These differences can be illustrated with copies of the Bible. While the Gospel According to Matthew nearly reached the practical limit of a scroll, a common codex included all four Gospels and the Book of Acts bound together, and complete Bibles were not uncommon.

The eventual supremacy of the codex form over scrolls and tablets was related to cultural and technological changes—i.e., the rise of Christianity, with its demand for more and larger books, and the availability of first parchment and then paper. The oldest extant Greek codex, said to date from the 4th century, is the Codex Sinaiticus, a biblical manuscript written in Greek. Also important is the Codex Alexandrinus, a Greek text of the Bible that probably was produced in the 5th century and is now preserved in the British Library, London.

In a completely separate development, codices also were made by the pre-Columbian peoples of Mesoamerica after about 1000 CE. These books contained pictographs and ideograms rather than written script. They dealt with the ritual calendar, divination, ceremonies, and speculations on the gods and the universe. Among these codices are the Vienna Codex, the Codex Colombino, and the Codex Fejérváry-Mayer, all believed to have been produced before the Spanish conquest of the region.

PAPER

Paper, which for more than 500 years has been the basic material used for written communication and the dissemination of information, is actually a matted or felted sheet of cellulose fibres formed from water suspension on a wire screen. The word *paper* is derived from the word "papyrus." However, complete defibring, an indispensable element in modern papermaking, did not occur in the preparation of papyrus sheets.

Modern paper is frequently traced to about 105 CE, when Cai Lun, an official attached to the Imperial court of China, is said to have created a sheet of paper using mulberry and other bast fibres along with fishnets, old rags, and hemp waste. The new material followed the caravan routes of Central Asia to the markets at Samarkand, from where it was distributed as a commodity across the entire Arab world. In 793 the first paper was made in Baghdad during the time of Hārūn ar-Rashīd, with the golden age of Islamic culture that brought papermaking to the frontiers of Europe.

The transmission of the techniques of papermaking appears to have followed the same route; Chinese taken prisoner at the Battle of Talas, near Samarkand, in 751 gave the secret to the Arabs. Paper mills proliferated from the end of the 8th century to the 13th century, from Baghdad and then on to Spain, then under Arab domination. Paper first penetrated Europe as a commodity from the 12th century onward through Italian ports that had active commercial relations with the Arab world and also, doubtless, by the overland route from Spain to France. Papermaking techniques apparently were rediscovered by Europeans through an examination of the material from which the imported commodity was made; possibly the secret was brought back in the mid-13th century by returning crusaders or merchants in the Eastern trade. Papermaking centres grew up in Italy after 1275 and in France and Germany during the 14th century.

The invention of printing in the 1450s brought a vastly increased demand for paper. Through the 18th century the papermaking process remained essentially unchanged, with linen and cotton rags furnishing the basic raw materials. Paper mills were increasingly plagued by shortages; in the 18th century they even advertised and solicited publicly for rags. It was evident that a process for utilizing a more abundant material was needed.

In 1800 a book was published that launched development of practical methods for manufacturing paper from wood pulp and other vegetable pulps. Several major pulping processes were gradually developed that relieved the paper industry of dependency upon cotton and linen rags and made modern large-scale production possible. These developments followed two distinct pathways. In one, fibres and fibre fragments were separated from the wood structure by mechanical means. In the other, the wood was exposed to chemical solutions that dissolved and removed lignin and other wood components, leaving cellulose fibre behind. Made by mechanical methods, groundwood pulp contains all the components of wood and thus is not suitable for papers in which high whiteness and permanence are required. Chemical wood pulps such as soda and sulfite pulp are used when high brightness, strength, and permanence are required. Groundwood pulp was first made in Germany in 1840, but the process did not come into extensive use until about 1870. Soda pulp was first manufactured from wood in 1852 in England, and in 1867 a patent was issued in the United States for the sulfite pulping process.

A sheet of paper composed only of cellulosic fibres ("waterleaf") is water absorbent. Hence, water-based inks and other aqueous liquids will penetrate and spread in it. Impregnation of the paper with various substances that retard such wetting and penetration is called sizing. Before 1800, paper sheets were sized by impregnation with animal glue or vegetable gums, an expensive and tedious process. In 1800 Moritz Friedrich Illig in Germany discovered that paper could be sized in vats with rosin and alum. Although Illig published his discovery in 1807, the method did not come into wide use for about 25 years.

Prior to the invention of the paper machine, paper was made one sheet at a time by dipping a frame or mold with a screened bottom into a vat of stock. Lifting the mold allowed the water to drain, leaving the sheet on the screen. The sheet was then pressed and dried. The size of a single sheet was limited to the size of frame and mold that a man could lift from a vat of stock. In 1798 Nicolas-Louis Robert in France constructed a moving screen belt that would receive a continuous flow of stock and deliver an unbroken sheet of wet paper to a pair of squeeze rolls. The French government recognized Robert's work by the granting of a patent. The paper machine did not become a practical reality, however, until two engineers in England, both familiar with Robert's ideas, built an improved version for their employers, Henry and Sealy Fourdrinier, in 1807. The Fourdrinier brothers obtained a patent also. Two years later a cylinder paper machine was devised by John Dickinson, an English papermaker. From these crude beginnings, modern papermaking machines evolved.

THE PRINTING PRESS

Few single inventions have had such far-reaching consequences as the 15th-century invention of printing with movable metal type. The details of this epochal invention are disappointingly obscure, but there is general agreement that the first large-scale printing workshop was established at Mainz, Ger., by Johannes Gutenberg, and was producing a sufficient quantity of accurate type to print a Vulgate Bible about 1455. Printers found an enormous demand for their product, so that the technique spread rapidly and the printed word became an essential medium of political, social, religious, and scientific communication as well as a convenient means for the dissemination of news and information. By 1500 almost 40,000 recorded editions of books had been printed in 14 European countries, with Germany and Italy accounting for two-thirds.

ORIGINS IN CHINA

By the end of the 2nd century CE, the Chinese apparently had discovered printing; certainly they then had at their disposal the three elements necessary for printing: (1) paper, the techniques for the manufacture of which they had known for several decades; (2) ink, whose basic formula they had known for 25 centuries; and (3) surfaces bearing texts carved in relief. Some of the texts were classics of Buddhist thought inscribed on marble pillars, to which pilgrims applied sheets of damp paper, daubing the surface with ink so that the parts that stood out in relief showed up; some were religious seals used to transfer pictures and texts of prayers to paper.

A substitute for these two kinds of surfaces, the marble pillars and the seals, appeared perhaps by the 6th century in the wood block. First, the text was written in ink on a sheet of fine paper. The written side of the sheet was then applied to the smooth surface of a block of wood and coated with a rice paste that retained the ink of the text. An engraver then cut away the uninked areas so that the text stood out in relief and in reverse. To make a print, the wood block was inked with a paintbrush, a sheet of paper spread on it, and the back of the sheet rubbed with a brush. Only one side of the sheet could be printed.

The oldest known printed works were made by this technique: in Japan about 764–770, Buddhist incantations ordered by Empress Shōtoku; in China in 868, the first known book, the Diamond Sūtra; and, beginning in 932, a collection of Chinese classics in 130 volumes, at the initiative of Fong Tao, a Chinese minister.

About 1041–48 a Chinese alchemist named Pi Sheng appears to have conceived of movable type made of an amalgam of clay and glue hardened by baking. It would thus appear that Pi Sheng had found an overall solution to the many problems of typography: the manufacture, the assembling, and the recovery of indefinitely reusable type. In about 1313 a magistrate named Wang Chen seems to have had a craftsman carve more than 60,000 characters on movable wooden blocks so that a treatise on the history of technology could be published. He is also credited with the invention of horizontal compartmented cases that revolved about a vertical axis to permit easier handling of the type. But Wang Chen's innovation, like that of Pi Sheng, was not followed up in China.

REINVENTION IN EUROPE

There is a material explanation for the fact that printing developed in Europe in the 15th century rather than in East Asia, even though the principle on which it is based had been known in the East long before. European writing was based on an alphabet composed of a limited number of abstract symbols. This simplified the problems involved in developing techniques for the use of movable type manufactured in series. Chinese handwriting, with its vast number of ideograms requiring some 80,000 symbols, lent itself only poorly to the requirements of a typography.

Xylography, the art of printing from wood carving, the existence and importance of which in China was never suspected by Marco Polo, appeared in Europe no earlier than the last quarter of the 14th century, spontaneously and presumably as a result of the use of paper. In the first half of the 15th century small, genuine books of several pages, religious works or compendiums of Latin grammar by Aelius Donatus and called *donats*, were published by a method identical to that of the Chinese.

Given the Western alphabet, it would seem reasonable that the next step taken might have been to carve blocks of writing that would simply contain a large number of letters of the alphabet; such blocks could then be cut up into type, usable and reusable. It is possible that experiments were in fact made along these lines, perhaps in 1423 or 1437 by a Dutchman from Haarlem, Laurens Janszoon, known as Coster. The encouraging results obtained with large type demonstrated the validity of the idea of typographic composition. However, cutting the small letters of the Roman alphabet from wood was a delicate operation. The type made in this way was fragile, and, since the letters were individually carved, no two copies of the same letter were identical. The process, thus, represented no advance in ease of production, durability, or quality.

Movable Type and the Printing Press

The association of die, matrix, and lead in the production of durable typefaces in large numbers and with each letter strictly identical was one of the two necessary elements in the invention of typographic printing in Europe. The second necessary element was the concept of the printing press itself, an idea that had never been conceived in the Far East.

Johannes Gutenberg is generally credited with the simultaneous discovery of both these elements, though there is some uncertainty about it, and disputes arose early to cloud the honour. Apart from chronicles, all published after his death, that attributed the invention of printing to him, probably the most convincing argument in favour of Gutenberg comes from his chief detractor, Johann Schöffer, the son of calligrapher Peter Schöffer and grandson of businessman Johann Fust, two men with whom Gutenberg had formed an association at Mainz, Ger. Though Schöffer claimed from 1509 on that the invention was solely his father's and grandfather's, he had, in 1505, written in a preface to an edition of Livy that "the admirable art of typography was invented by the ingenious Johan Gutenberg at Mainz in 1450."

The first pieces of type appear to have been made in the following steps: a letter die was carved in a soft metal such as brass or bronze; lead was poured around the die to form a matrix and a mold into which an alloy, which was to form the type itself, was poured. It was probably Peter Schöffer who, around 1475, thought of replacing the soft-metal dies with steel dies, in order to produce copper letter

THE GUTENBERG BIBLE

The first complete book extant in the West and the earliest printed from movable type is a Vulgate Bible (that is, a Bible in Latin translation) named after its printer, Johannes Gutenberg, who completed it about 1455 working at Mainz, Ger. The three-volume work was printed in 42-line columns and, in its later stages of production, was worked on by six compositors simultaneously. It is sometimes referred to as the Mazarin Bible because the first copy described by bibliographers was located in the Paris library of Cardinal Mazarin.

A page from the 42-line Gutenberg Bible, 1455. Courtesy of the Newberry Library, Chicago

The 42-line Gutenberg Bible, printed in Mainz, Ger., in 1455. Rare Books and Manuscripts Division, the New York Public Library; Astor, Lenox, and Tilden Foundations

Like other contemporary works, the Gutenberg Bible had no title page, no page numbers, and no innovations to distinguish it from the work of a manuscript copyist. This was presumably the desire of both Gutenberg and his customers. Experts generally agree that the Bible, though uneconomic in its use of space, displays a technical efficiency not substantially improved upon before the 19th century. The Gothic type is majestic in appearance, medieval in feeling, and slightly less compressed and less pointed than other examples that appeared shortly thereafter.

The original number of copies of this work is unknown; some 40 are still in existence. There are perfect vellum copies in the U.S. Library of Congress, the French Bibliotheque Nationale, the British Library, and the Göttingen State and University Library in Germany. In the United States almost-complete texts are in the Huntington, Morgan, New York Public, Harvard University, and Yale University libraries.

matrices that would be reliably identical. Until the middle of the 19th century, type generally continued to be made by craftsmen in this way.

Documents of the period, including those relating to a 1439 lawsuit in connection with Gutenberg's activities at Strassburg, leave little doubt that the press has been used since the beginning of printing. Perhaps the printing press was first just a simple adaptation of the binding press, with a fixed, level lower surface (the bed) and a movable, level upper surface (the platen), moved vertically by means of a small bar on a worm screw. The composed type, after being locked by ligatures or screwed tight into a right metal frame (the form), was inked, covered with a sheet of paper to be printed, and then the whole pressed in the vise formed by the two surfaces.

This process was superior to the brushing technique used in wood-block printing in Europe and China because it was possible to obtain a sharp impression and to print both sides of a sheet. Nevertheless, there were deficiencies. It was difficult to pass the leather pad used for inking between the platen and the form; and, since several turns of the screw were necessary to exert the required pressure, the bar had to be removed and replaced several times to raise the platen sufficiently to insert the sheet of paper.

It is generally thought that the printing press acquired its principal functional characteristics very early, probably before 1470. The first of these may have been the mobile bed, either on runners or on a sliding mechanism, that permitted the form to be withdrawn and inked after each sheet was printed. Next, the single thread of the worm screw was replaced with three or four parallel threads with a sharply inclined pitch so that the platen could be raised by a slight movement of the bar. This resulted in a decrease in the pressure exerted by the platen, which was corrected by breaking up the printing operation so that the form was pushed under the press by the movable bed so that first one half and then the other half of the form was utilized. This was the principle of printing "in two turns," which would remain in use for three centuries.

PHOTOGRAPHY

Photography has aptly been called the most important invention since the printing press. An effective photograph can disseminate information about humanity and nature, record the visible world, and extend human knowledge and understanding. One characteristic is unique to photography and sets it apart from other ways of picture making: the essential components of the image are usually established immediately at the time of exposure to the light-sensitive elements of the camera. This seemingly automatic recording of an image by photography has given the process a sense of authenticity shared by no other picture-making technique. The photograph possesses, in the popular mind, such apparent accuracy that the adage "the camera does not lie" has become a widely accepted cliché.

Antecedents

The forerunner of the camera was the camera obscura, a dark chamber or room with a hole (later a lens) in one wall, through which images of objects outside the room were projected on the opposite wall. The principle was probably known to the Chinese and to ancient Greeks such as Aristotle more than 2,000 years ago. Late in the 16th century, the Italian scientist and writer Giambattista della Porta demonstrated and described in detail the use of a camera obscura with a lens. While artists in subsequent centuries commonly used variations on the camera obscura to create images they could trace, the results from these devices depended on the artist's drawing skills, and so scientists continued to search for a method to reproduce images completely mechanically.

In 1727 the German professor of anatomy Johann Heinrich Schulze proved that the darkening of silver salts, a phenomenon known since the 16th century and possibly earlier, was caused by light and not heat. He demonstrated the fact by using sunlight to record words on the salts, but he made no attempt to preserve the images permanently. His discovery, in combination with the camera obscura, provided the basic technology necessary for photography. It was not until the early 19th century, however, that photography actually came into being.

Heliography

Nicéphore Niépce, an amateur inventor living near Chalon-sur-Saône, a city 189 miles (304 km) southeast of Paris, was interested in lithography, a process in which drawings are copied or drawn by hand onto lithographic stone and then printed in ink. Not artistically trained, Niépce devised a method by which light could draw the pictures he needed. He oiled an engraving to make it transparent, placed it on a plate coated with a light-sensitive solution of bitumen of Judea (a type of asphalt) and lavender oil, and exposed the setup to sunlight. After a few hours, the solution under the light areas of the engraving hardened, while that under the dark areas remained soft and could be washed away, leaving a permanent, accurate copy of the engraving. Calling the process heliography ("sun drawing"), Niépce succeeded from 1822 onward in copying oiled engravings onto lithographic stone, glass, and zinc, and from 1826 onto pewter plates.

In 1826–27, using a camera obscura fitted with a pewter plate, Niépce produced the first successful photograph from nature, a view of the courtyard of his country estate, Gras, from an upper window of the house. The exposure time was about eight hours, during which the sun moved from east to west so that it appears to shine on both sides of the building.

Niépce produced his most successful copy of an engraving, a portrait of Cardinal d'Amboise, in 1826. It was exposed in about three hours, and in February 1827 he had the pewter plate etched to form a printing plate and had two prints pulled. Paper prints were the final aim of Niépce's heliographic process, yet all his other attempts, whether made by using a camera or by means of engravings, were underexposed and too weak to be etched. Nevertheless, Niépce's discoveries showed the path that others were to follow with more success.

Daguerreotype

Louis-Jacques-Mandé Daguerre was a professional scene painter for the theatre. Between 1822 and 1839 he was coproprietor of the Diorama in Paris, an auditorium in which he and his partner Charles-Marie Bouton displayed immense paintings, 45.5 by 71.5 feet (14 by 22 m) in size, of famous places and historical events. The partners painted the scenes on translucent paper or muslin and, by the careful use of changing lighting effects, were able to present vividly realistic tableaux. The views provided grand, illusionistic entertainment, and the amazing trompe l'oeil effect was purposely heightened by the accompaniment of appropriate music and the positioning of real objects, animals, or people in front of the painted scenery.

Like many other artists of his time, Daguerre made preliminary sketches by tracing the images produced by both the camera obscura and the camera lucida, a prism-fitted instrument that was invented in 1807. His attempt to retain the duplication of nature he perceived in the camera obscura's ground glass led in 1829 to a partnership with Niépce, with whom he worked in person and by correspondence for the next four years. However, Daguerre's interest was in shortening the exposure time necessary to obtain an image of the real world, while Niépce remained interested in producing reproducible plates. It appears that by 1835, three years after Niépce's death, Daguerre had discovered that a latent image forms on a plate of iodized silver and that it can be "developed" and made visible by exposure to mercury vapour, which settles on the exposed parts of the image. Exposure times could thus be reduced from eight hours to 30 minutes. The results were not permanent, however; when the developed picture was exposed to light, the unexposed areas of silver

GEORGE EASTMAN

American entrepreneur and inventor George Eastman was born on July 12, 1854, in Waterville, N.Y. After his education in the public schools of Rochester, N.Y., Eastman worked briefly for an insurance company and a bank. In 1880 he perfected a process of making dry plates for photography and organized the Eastman Dry Plate and Film Company for their manufacture. The first Kodak (a name he coined) camera was placed on the market in 1888. It was a simple, handheld box camera containing a 100-exposure roll of film that used paper negatives. Consumers sent the entire camera back to the manufacturer for developing, printing, and reloading when the film was used up: the company's slogan was "You press the button, we do the rest." In 1889 Eastman introduced roll film on a transparent base, which has remained the standard for film. In 1892 he reorganized the business as the Eastman Kodak Company. Eight years later he introduced the Brownie camera, which was intended for use by children and sold for $1. By 1927 Eastman Kodak had a virtual monopoly of the photographic industry in the United States, and it has continued to be one of the largest American companies in its field.

Eastman gave away half his fortune in 1924. His gifts, which totaled more than $75 million, went to such beneficiaries as the University of Rochester (of which the Eastman School of Music is a part) and the Massachusetts Institute of Technology in Boston. He was also one of the first owners to introduce profit sharing as an employee incentive. Eastman took his own life at home on March 14, 1932, at age 77, leaving a note that said, "My work is done. Why wait?" His home in Rochester, now known as George Eastman House, has become a renowned archive and museum of international photography as well as a popular tourist site.

darkened until the image was no longer visible. By 1837 Daguerre was able to fix the image permanently by using a solution of table salt to dissolve the unexposed silver iodide. That year he produced a photograph of his studio on a silvered copper plate, a photograph that was remarkable for its fidelity and detail. Also that year, Niépce's son Isidore signed an agreement with Daguerre affirming Daguerre as the inventor of a new process, "the daguerreotype."

In 1839 Niépce's son and Daguerre sold full rights to the daguerreotype and the heliograph to the French government, in return for annuities for life. On August 19 of that year, full working details were published. Daguerre wrote a booklet describing the process, *An Historical and Descriptive Account of the Various Processes of the Daguerreotype and the Diorama*, which at once became a best seller; 29 editions and translations appeared before the end of 1839.

THE TELEGRAPH

The word *telegraph* is derived from the Greek words *tele*, meaning "distant," and *graphein*, meaning "to write." It came into use toward the end of the 18th century to describe an optical semaphore system developed in France, but the term is most often understood to refer to the electric telegraph, which was developed in the mid-19th century and for more than 100 years was the principal means of transmitting printed information by wire or radio wave.

PREELECTRIC TELEGRAPH SYSTEMS

Before the development of the electric telegraph, visual systems were used to convey messages over distances by means of variable displays. One of the most successful of the visual telegraphs was the semaphore developed in France by the Chappe brothers, Claude and Ignace, in 1791.

This system consisted of pairs of movable arms mounted at the ends of a crossbeam on hilltop towers. Each arm of the semaphore could assume seven angular positions 45° apart, and the horizontal beam could tilt 45° clockwise or counterclockwise. In this manner it was possible to represent numbers and the letters of the alphabet. Chains of these towers were built to permit transmission over long distances. The towers were spaced at intervals of 3 to 6 miles (5 to 10 km), and a signaling rate of three symbols per minute could be achieved. The two-flag semaphore system was widely used well into the 20th century, particularly by the world's navies.

Another widely used visual telegraph was developed in 1795 by George Murray in England. In Murray's device, characters were sent by opening and closing various combinations of six shutters. This system rapidly caught on in England and in the United States, where a number of sites bearing the name Telegraph Hill or Signal Hill can still be found, particularly in coastal regions. Visual telegraphs were completely replaced by the electric telegraph by the middle of the 19th century.

The Morse System

The electric telegraph did not burst suddenly upon the scene but rather resulted from a scientific evolution that had been taking place since the 18th century in the field of electricity. One of the key developments was the invention of the voltaic cell in 1800 by Alessandro Volta of Italy. This made it possible to power electric devices in a more effective manner using relatively low voltages and high currents. Previous methods of producing electricity employed frictional generation of static electricity, which led to high voltages and low currents. Many devices incorporating high-voltage static electricity and various detectors such as pith balls and sparks were proposed for use in telegraphic systems. All were unsuccessful, however, because the severe losses in the transmission wires, particularly in bad weather, limited reliable operation to relatively short distances. Application of the battery to telegraphy was made possible by several further developments in the new science of electromagnetism. In 1820 Hans Christian Ørsted of Denmark discovered that a magnetic needle could be deflected by a wire carrying an electric current. In 1825 in Britain William Sturgeon discovered the multiturn electromagnet, and in 1831 Michael Faraday of Britain and Joseph Henry of the United States refined the science of electromagnetism sufficiently to make it possible to design practical electromagnetic devices.

The first two practical electric telegraphs appeared at almost the same time. In 1837 the British inventors Sir William Fothergill Cooke and Sir Charles Wheatstone obtained a patent on a telegraph system that employed six wires and actuated five needle pointers attached to five galvanoscopes at the receiver. If currents were sent through the proper wires, the needles could be made to point to specific letters and numbers on their mounting plate.

In 1832 Samuel F. B. Morse, a professor of painting and sculpture at the University of the City of New York (later New York University), became interested in the possibility of electric telegraphy and made sketches of ideas for such a system. In 1835 he devised a system of dots and dashes to represent letters and numbers. In 1837 he was granted a patent on an electromagnetic telegraph. Morse's original transmitter incorporated a device called a portarule, which employed molded type with built-in dots and dashes. The type could be moved through a mechanism in such a manner that the dots and dashes would make and break the contact between the battery and the wire to the receiver. The receiver, or register, embossed the dots and dashes on an unwinding strip of paper that passed under a stylus. The stylus was actuated by an electromagnet turned on and off by the signals from the transmitter.

Morse had formed a partnership with Alfred Vail, who was a clever mechanic and is credited with many contributions to the Morse system. Among them are the replacement of the portarule transmitter by a simple make-and-break key, the refinement of the Morse code so that the shortest code sequences were assigned to the most frequently occurring letters, and the improvement of the mechanical design of all the system components. The first demonstration of the system by Morse was conducted for his friends at his workplace in 1837. In 1843 Morse obtained financial support from the U.S. government to build a demonstration telegraph system 35 miles (60 km) long between Washington, D.C., and Baltimore, Md. Wires were attached by glass insulators to poles alongside a railroad. The system was completed and public use initiated on May 24, 1844, with transmission of the message, "What hath God wrought!" This inaugurated the telegraph era in the United States, which was to last more than 100 years.

THE TELEPHONE

The word *telephone*, from the Greek roots *tēle*, "far," and *phonē*, "sound," was applied as early as the late 17th century to the string telephone familiar to children, and it was later used to refer to the megaphone and the speaking tube, but in modern usage it refers solely to electrical devices derived from the inventions of Alexander Graham Bell and others. Within 20 years of the 1876 Bell patent, the telephone instrument, as modified by Thomas Watson, Emil Berliner, Thomas Edison, and others, acquired a functional design that has not changed fundamentally in more than a century. Since the invention of the transistor in 1947, metal wiring and other heavy hardware have been replaced by lightweight and compact microcircuitry. Advances in electronics have improved the performance of the basic design, and they also have allowed the introduction of a number of "smart" features such as automatic redialing, call-number identification, wireless transmission, and visual data display. Such advances supplement, but do not replace, the basic telephone design conceived by Bell in the 1870s.

Early Sound Transmitters

Beginning in the early 19th century, several inventors made a number of attempts to transmit sound by electric means. The first inventor to suggest that sound could be transmitted electrically was a Frenchman, Charles Bourseul, who indicated that a diaphragm making and breaking contact with an electrode might be used for this purpose. By 1861 Johann Philipp Reis of Germany had designed several instruments for the transmission of sound. The transmitter Reis employed consisted of a membrane with a metallic strip that would intermittently contact a metallic point connected to an electrical circuit. As sound waves impinged on the membrane, making the membrane vibrate, the circuit would be connected and interrupted at the same rate as the frequency of the sound. The fluctuating electric current thus generated would be transmitted by wire to a receiver, which consisted of an iron needle that was surrounded by the coil of an electromagnet and connected to a sounding box. The fluctuating electric current would generate varying magnetic fields in the coil, and these in turn would force the iron needle to produce vibrations in the sounding box. Reis's system could thus transmit a simple tone, but it could not reproduce the complex waveforms that make up speech.

Gray and Bell: the Transmission of Speech

The First Devices

In the 1870s two American inventors, Elisha Gray and Alexander Graham Bell, each independently designed devices that could transmit speech electrically. Gray's first device made use of a harmonic telegraph, the transmitter and receiver of which consisted of a set of metallic reeds tuned to different frequencies. An electromagnetic coil was located near each of the reeds. When a reed in the transmitter was vibrated by sound waves of its resonant frequency—for example, 400 hertz—it induced an electric

Alexander Graham Bell demonstrating the ability of the telephone to transmit sound by electricity from Salem to Boston, Mass., 1887. Library of Congress, Washington, D.C.

current of corresponding frequency in its matching coil. This coil was connected to all the coils in the receiver, but only the reed tuned to the transmitting reed's frequency would vibrate in response to the electric current. Thus, simple tones could be transmitted. In the spring of 1874 Gray realized that a receiver consisting of a single steel diaphragm in front of an electromagnet could reproduce any of the transmitted tones. Gray, however, was initially unable to conceive of a transmitter that would transmit complex speech vibrations and instead chose to demonstrate the transmission of tones via his telegraphic device in the summer of 1874.

Bell, meanwhile, also had considered the transmission of speech using the harmonic telegraph concept, and in the summer of 1874 he conceived of a membrane receiver similar to Gray's. However, since Bell, too, had no transmitter, the membrane device was never constructed. Following some earlier experiments, Bell postulated that, if two membrane receivers were connected electrically, a sound wave that caused one membrane to vibrate would induce a voltage in the electromagnetic coil that would in turn cause the other membrane to vibrate. Working with a young machinist, Thomas Augustus Watson, Bell had two such instruments constructed in June 1875. The device was tested on June 3, 1875, and, although no intelligible words were transmitted, "speechlike" sounds were heard at the receiving end.

An application for a U.S. patent on Bell's work was filed on Feb. 14, 1876. Several hours later that same day, Gray filed a caveat on the concept of a telephone transmitter and receiver. A caveat was a confidential, formal declaration by an inventor to the U.S. Patent Office of an intent to file a patent on an idea yet to be perfected; it was intended to prevent the idea from being used by other inventors. At this point neither Gray nor Bell had yet

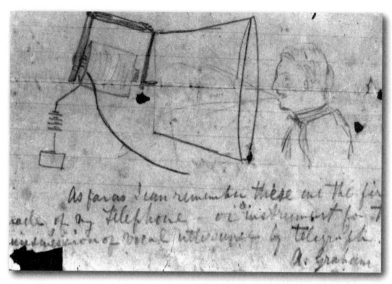

Alexander Graham Bell's sketch of a telephone. He filed the patent for his telephone at the U.S. Patent Office on Feb. 14, 1876—just hours before a rival, Elisha Gray, filed a declaration of intent to file a patent for a similar device. © Photos.com/Jupiterimages

constructed a working telephone that could convey speech. On the basis of its earlier filing time, Bell's patent application was allowed over Gray's caveat. On March 7, 1876, Bell was awarded U.S. patent 174,465. This patent is often referred to as the most valuable ever issued by the U.S. Patent Office, as it described not only the telephone instrument but also the concept of a telephone system.

THE SEARCH FOR A SUCCESSFUL TRANSMITTER

Gray had earlier come up with an idea for a transmitter in which a moving membrane was attached to an electrically conductive rod immersed in an acidic solution. Another conductive rod was immersed in the solution, and, as sound waves impinged on the membrane, the two rods would move with respect to each other. Variations in the distance between the two rods would produce variations in electric resistance and, hence, variations in the electric current. In

Communication

Alexander Graham Bell filing the patent for his telephone at the U.S. Patent Office on Feb. 14, 1876. Bell's telephone is on the table to the right. © Photos.com/Jupiterimages

contrast to the magnetic coil type of transmitter, the variable-resistance transmitter could actually amplify the transmitted sound, permitting use of longer cables between the transmitter and the receiver.

Again, Bell also worked on a similar "liquid" transmitter design; it was this design that permitted the first transmission of speech, on March 10, 1876, by Bell to Watson: "Mr. Watson, come here. I want you." The first public demonstrations of the telephone followed shortly

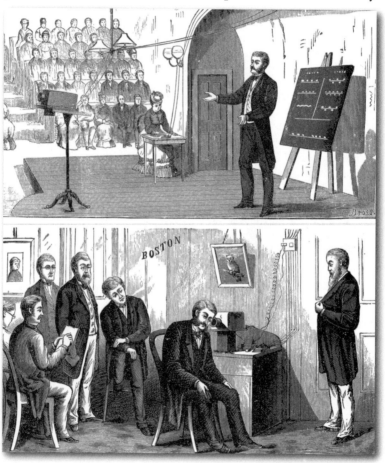

Alexander Graham Bell, inventor who patented the telephone in 1876, lecturing at Salem, Mass. (top), while friends in his study at Boston listened to his lecture via telephone, Feb. 12, 1877. © Photos.com/Jupiterimages

afterward, featuring a design similar to the earlier magnetic coil membrane units described above. One of the earliest demonstrations occurred in June 1876 at the Centennial Exposition in Philadelphia. Further tests and refinement of equipment followed shortly afterward. On Oct. 9, 1876 Bell conducted a two-way test of his telephone over a

THE AT&T DESK PHONE

The earliest telephone instrument to see common use was introduced by Charles Williams Jr. in 1882. Designed for wall mounting, this instrument consisted of a ringer, a hand-cranked magneto (for generating a ringing voltage in a distant instrument), a hand receiver, a switch hook, and a transmitter. Various versions of this telephone instrument remained in use throughout the United States as late as the 1950s. Telephone dial originated with automatic telephone switching systems in 1896.

Desk instruments were first constructed in 1897. Patterned after the wall-mounted telephone, they usually consisted of a separate receiver and transmitter. In 1927, however, the American Telephone & Telegraph Company (AT&T) introduced the E1A handset, which employed a combined transmitter-receiver arrangement. The ringer and much of the telephone electronics remained in a separate box, on which the transmitter-receiver handle was cradled when not in use. The first telephone to incorporate all the components of the station apparatus into one instrument was the so-called combined set of 1937. Some 25 million of these instruments were produced until they were superseded by a new design in 1949. The 1949 telephone was totally new, incorporating significant improvements in audio quality, mechanical design, and physical construction. Push-button versions of this set became available in 1963.

2-mile (5-km) distance between Boston and Cambridgeport, Mass. In May 1877 the first commercial application of the telephone took place with the installation of telephones in offices of customers of the E.T. Holmes burglar alarm company.

The poor performance of early telephone transmitters prompted a number of inventors to pursue further work in this area. Among them was Thomas Alva Edison, whose 1886 design for a voice transmitter consisted of a cavity filled with granules of carbonized anthracite coal. The carbon granules were confined between two electrodes through which a constant electric current was passed. One of the electrodes was attached to a thin iron diaphragm, and, as sound waves forced the diaphragm to vibrate, the carbon granules were alternately compressed and released. As the distance across the granules fluctuated, resistance to the electric current also fluctuated, and the resulting variations in current were transmitted to the receiver. Edison's carbon transmitter was so simple, effective, cheap, and durable that it became the basis for standard telephone transmitter design through the 1970s.

THE PHONOGRAPH

On July 18, 1877, the American inventor Thomas Edison was working in his Menlo Park, N.J., laboratory, trying to improve upon the telephone of Alexander Graham Bell. Thinking he might find a way to record telephone messages for relay over long distances, he attached a hard tip to a telephone diaphragm and spoke the words "Halloo! Halloo!" into the mouthpiece while passing a strip of waxed paper through the device. When the paper strip was pulled back over the diaphragm, he was amazed to hear the words played back to him. That day can be said to mark the beginning of practical sound recording.

Edison's Phonograph

Twenty years earlier the French printer Édouard-Léon Scott de Martinville had made a "phonoautograph" that traced sound waves on graphing paper, and only a few months previously, in April 1877, the French poet Charles Cros had described a process for recording sound waves on a glass disc. Yet despite such progress, nobody had made a workable sound-recording apparatus until Edison. Edison followed his initial discovery by designing and patenting a device with a vibrating membrane attached to a rigid stylus that indented the tinfoil surface of a hand-cranked grooved cylinder. Upon being retraced by a lighter stylus, the indentations were sufficient to reproduce the original vibrations. The phonograph, as Edison called it, was the first practical machine to record and play back speech and music. The press immediately praised the miracle of recorded sound.

The phonograph worked, but it did not sell. Although Edison considered it one of his most original inventions, he was unable at first to improve its poor sound quality or mass-produce the recordings. The cylinders, which he called "phonograms," were standardized at 4 inches (10 cm) long and $2^{1}/_{4}$ inches (6 cm) in diameter, with 100 grooves per inch, but they played for only about one minute and wore out quickly. The Edison Speaking Phonograph Company sold approximately 2,000 machines before closing down production in 1878.

The Graphophone and Gramophone

During this period two improved recording machines were introduced by other inventors. First was the graphophone, patented in 1886 by Charles Sumner Tainter, Chichester Bell, and Alexander Graham Bell of the Volta

Laboratory in Washington, D.C. The graphophone used a smaller-diameter cylinder coated with wax rather than tinfoil. The softer recording surface made it possible to cut smaller "hill-and-dale" vertical grooves in the cylinder (as opposed to the indentations made by the Edison machine), allowing a longer playing time of two minutes.

The second machine was the 1888 gramophone of Emil Berliner, a German immigrant also working in Washington. Berliner's instrument recorded sound on a flat 5-inch (13-cm) wax-coated zinc disc rather than cylinders. The recording stylus moved laterally when cutting into the thin wax layer, tracing a zigzag groove that produced louder volume during playback. The playback disc itself was produced by pressing a suitable molding material such as hard rubber or celluloid against the zinc master.

The Race to Market

The inventors were not unaware of the commercial potential of their inventions. By the time they obtained their patents, Tainter and the Bells had already founded the Volta Graphophone Company, and Berliner established the U.S. Gramophone Company in 1893. As early as 1888, Berliner predicted that discs would become the medium of choice for consumers; however, it was cylinders that enjoyed the first success.

Edison returned to his phonograph in 1886 and produced a new model with a sapphire floating stylus cutting a hill-and-dale groove only $1/1000$-inch (0.025 mm) deep in a cylinder of solid white wax. His 1888 Improved Phonograph replaced the hand crank of his first machine with an electric motor that ran at 125 rotations per minute. New 6-inch (15-cm) cylinders played for almost two minutes.

In 1888 the patent rights of the Edison and Graphophone companies were purchased by entrepreneur Jesse

Thomas Alva Edison demonstrating his tinfoil phonograph. Photograph by Mathew Brady, 1878. Courtesy of the Edison National Historical Site, West Orange, N.J.

Lippincott, who founded the North American Phonograph Company to lease dictating machines to local distributors. However, stenographers refused to use the machines, and Lippincott's company failed by 1894. Meanwhile, one of the original graphophone investors, Edward D. Easton, declined to sell his interests to Lippincott and instead started the Columbia Phonograph Company. Easton specialized in popular music recordings, featuring such attractions as John Philip Sousa's Marine Band. But no national market for recorded music had yet developed.

COIN SLOTS AND HOME PHONOGRAPHS

It was in the world of public entertainment that sound recording found its first widespread acceptance. In 1889, Louis Glass of San Francisco invented a mechanism to start the motor of an electric phonograph when a nickel was deposited into a slot. Within a year, he joined Felix Gottschalk to create the Automatic Phonograph Company, which placed phonographs in all types of entertainment establishments, from saloons to ice cream parlors. The talking machine thus joined the American mass leisure culture that emerged in the 1890s, along with Coney Island, professional baseball, vaudeville, and movie theatres.

The coin-slot business energized the manufacturing companies. Easton's Columbia acquired the Tainter-Bell patents from Lippincott's failed company and in 1894 introduced a low-cost graphophone with a spring motor developed by Thomas H. Macdonald. For the first time, the graphophone could play Edison standard cylinders and was priced for the masses. The subsequent competition between Edison and Macdonald for the new emerging home market was intense. Columbia focused its sales effort on big cities, while Edison appealed more to rural buyers. Edison recovered his patent rights from Lippincott and organized the National Phonograph Company in 1896 to sell a $40. Home Phonograph, touted as the "machine for the millions." When Columbia lowered its price to $25, Edison improved his Home model's spring motor in 1897 to play six cylinders with one winding and lowered his price to $30.

Meanwhile, Berliner finally developed a method to mass-produce his flat discs, stamping rubber copies from a hardened electroplated master that did not wear out like the wax masters. Berliner then hired Eldridge Johnson in 1896 to make a spring motor for his $25. Improved Gramophone and to improve the duplication process. The result was a sound of clearly improved quality. Johnson used his patents and those of Berliner in 1901 to create a new company, the Victor Talking Machine Company, which shared the famous Nipper dog trademark first acquired by the British Gramophone Company. Within a year Victor established a national network of 5,000 sales outlets; Johnson also launched an advertising campaign in the mass media to promote the gramophone as affordable to everyone.

Thus by the beginning of the new century, sound recording had grown into a mass medium owing to a few key factors: visionary inventors such as Eldridge Johnson and Emile Berliner, the coin-slot business, the low-cost spring motor player, the electroplate duplication method, and aggressive marketing techniques. The war between cylinders and discs started by Berliner in 1888 was finally won by the disc 25 years later. The flat disc was easier to use and store, and Eldridge Johnson, head of the Victor Talking Machine Company, was a superior promoter. In 1902 Victor and Columbia agreed to share patents. Columbia began to produce disc players immediately after, and in 1908 the company stopped making cylinders altogether. Edison finally conceded that the disc had won and introduced the Diamond Disc flat record and player in 1913.

MOTION PICTURES

Motion-picture photography is based on the phenomenon that the human brain will perceive an illusion of continuous movement from a succession of still images exposed at a rate above 15 frames per second. Although posed sequential pictures had been taken as early as 1860, successive photography of actual movement was not achieved until 1877, when Eadweard Muybridge used 12 equally spaced cameras to demonstrate that at some time all four hooves of a galloping horse left the ground at once. In 1877–78 an associate of Muybridge devised a system of magnetic releases to trigger an expanded battery of 24 cameras.

The Muybridge pictures were widely published in still form. They were also made up as strips for the popular parlour toy the zoetrope "wheel of life," a rotating drum that induced an illusion of movement from drawn or painted pictures. Meanwhile, Émile Reynaud in France was projecting sequences of drawn pictures onto a screen using his Praxinoscope, in which revolving mirrors and an oil-lamp "magic lantern" were applied to a zoetrope-like drum. By 1880 Muybridge was similarly projecting enlarged, illuminated views of his motion photographs using the Zoöpraxiscope, an adaptation of the zoetrope.

SEQUENTIAL PHOTOGRAPHS

Although a contemporary observer of Muybridge's demonstration claimed to have seen "living, moving animals," such devices lacked several essentials of true motion pictures. The first was a mechanism to enable sequence photographs to be taken within a single camera at regular, rapid intervals, and the second was a medium capable of storing images for more than the second or so of movement possible from drums, wheels, or disks.

A motion-picture camera must be able to advance the medium rapidly enough to permit at least 16 separate exposures per second as well as bring each frame to a full stop to record a sharp image. The principal technology that creates this intermittent movement is the Geneva watch movement, in which a four-slotted star wheel, or "Maltese cross," converts the tension of the mainspring to the ticking of toothed gears. In 1882 Étienne-Jules Marey employed a similar "clockwork train" intermittent movement in a photographic "gun" used to "shoot" birds in flight. Twelve shots per second could be recorded onto a circular glass plate. Marey subsequently increased the frame rate, although for no more than about 30 images, and employed strips of sensitized paper (1887) and paper-backed celluloid (1889) instead of the fragile, bulky glass. The transparent material trade-named celluloid was first manufactured commercially in 1872. It was derived from collodion, that is, nitrocellulose (gun cotton) dissolved in alcohol and dried. John Carbutt manufactured the first commercially successful celluloid photographic film in 1888, but it was too stiff for convenient use. By 1889 the George Eastman company had developed a roll film of celluloid coated with photographic emulsion for use in its Kodak still camera. This sturdy, flexible medium could transport a rapid succession of numerous images and was eventually adapted for motion pictures.

Edison's Kinetograph and Kinetoscope

Thomas Edison is often credited with the invention of the motion picture in 1889. The claim is disputable, however, specifically because Edison's motion-picture operations were entrusted to an assistant, W. K. L. Dickson, and generally because there are several plausible pre-Edison claimants in England and France. Indeed, a U.S. Supreme Court decision of 1902 concluded that Edison had not invented the motion picture but had only combined the discoveries of others. His systems are important, nevertheless, because they prevailed commercially. The heart of Edison's patent claim was the intermittent movement provided by a Maltese cross synchronized with a shutter. The October 1892 version of Edison's Kinetograph camera employed the format essentially still in use today. The film, made by Eastman according to Edison's specifications, was 35 millimetres (mm) in width. Two rows of sprocket holes, each with four holes per frame, ran the length of the film and were used to advance it. The image was 1 inch (2.5 cm) wide by ¾ inch (1.9 cm) high.

At first Edison's motion pictures were not projected. One viewer at a time could watch a film by looking through the eyepiece of a peep-show cabinet known as the Kinetoscope. This device was mechanically derived from the zoetrope in that the film was advanced by continuous movement, and action was "stopped" by a very brief exposure. In the zoetrope, a slit opposite the picture produced a stroboscopic effect; in the Kinetoscope the film traveled at the rate of 40 frames per second, and a slit in a 10-inch (25 cm) diameter rotating shutter wheel afforded an exposure of 6,000 seconds. Illumination was provided by an electric bulb positioned directly beneath the film. The film ran over spools. Its ends were spliced together to form a continuous loop, which was initially 25 to 30 feet (7 to 9 m) long but later was lengthened to almost 50 metres. A direct-current motor powered by an Edison storage battery moved the film at a uniform rate.

The Kinetoscope launched the motion-picture industry, but its technical limitations made it unsuitable for projection. Films may run continuously when a great deal of light is not crucial, but a bright, enlarged picture requires that each frame be arrested and exposed intermittently as in the camera. The adaptation of the camera mechanism to projection seems obvious in retrospect but was frustrated in the United States by Dickson's establishment of a frame rate well above that necessary for the perception of continuous motion.

THE *CINÉMATOGRAPHE*

After the Kinetoscope was introduced in Paris, Auguste and Louis Lumière produced a combination camera/projector, first demonstrated publicly in 1895 and called the *cinématographe*. The device used a triangular "eccentric" (intermittent) movement connected to a claw to engage the sprocket holes. As the film was stationary in the aperture for two-thirds of each cycle, the speed of 16 frames per second allowed an exposure of $^1/_{25}$ second. At this slower rate audiences could actually see the shutter blade crossing the screen, producing a "flicker" that had been absent from Edison's pictures. On the other hand, the hand-cranked *cinématographe* weighed less than 20 pounds (9 kg) (Edison's camera weighed 100 times as much). The Lumière units could therefore travel the world to shoot and screen their footage. The first American projectors employing intermittent movement were devised by Thomas Armat in 1895 with a Pitman arm or "beater" movement taken from a French camera of 1893. The following year Armat agreed to allow Edison to produce the projectors in quantity and to market them as Edison Vitascopes. In 1897 Armat patented the first projector with four-slot star and cam (as in the Edison camera).

One limitation of early motion-picture filming was the tearing of sprocket holes. The eventual solution to this problem was the addition to the film path of a slack-forming loop that restrained the inertia of the take-up reel. When this so-called Latham loop was applied to cameras and projectors with intermittent movement, the growth and shrinkage of the loops on either side of the shutter adjusted for the disparity between the stop-and-go motion at the aperture and the continuous movement of the reels.

When the art of projection was established, the importance of a bright screen picture was appreciated. Illumination was provided by carbon arc lamps, although flasks of ether and sticks of unslaked calcium ("limelight") were used for brief runs.

RADIO

Early in the 19th century, Michael Faraday, an English physicist, demonstrated that an electric current can produce a local magnetic field and that the energy in this field will return to the circuit when the current is stopped or changed. James Clerk Maxwell, professor of experimental physics at Cambridge, in 1864 proved mathematically that any electrical disturbance could produce an effect at a considerable distance from the point at which it occurred and predicted that electromagnetic energy could travel outward from a source as waves moving at the speed of light. The scientific examination of the relationship between light waves and electromagnetic waves thus revealed the possibility of transmitting electromagnetic signals between widely separated points, preparing the way for Italian physicist Guglielmo Marconi to transmit the first wireless message across the Atlantic on Dec. 12, 1901.

HERTZ: RADIO-WAVE EXPERIMENTS

At the time of Maxwell's prediction there were no known means of propagating or detecting the presence of electromagnetic waves in space. It was not until about 1888 that Maxwell's theory was tested by Heinrich Hertz, who demonstrated that Maxwell's predictions were true at least over short distances by installing a spark gap (two conductors separated by a short gap) at the centre of a parabolic metal mirror. A wire ring connected to another spark gap was placed about 5 feet (1.5 m) away at the focus of another parabolic collector in line with the first. A spark jumping across the first gap caused a smaller spark to jump across the gap in the ring 5 feet away. Hertz showed that the waves travelled in straight lines and that they could be reflected by a metal sheet just as light waves are reflected by a mirror.

Marconi's Wireless Telegraph

Marconi, whose main genius was in his perseverance and refusal to accept expert opinion, repeated Hertz's experiments and eventually succeeded in getting secondary sparks over a distance of 30 feet (9 m). In his experiment he attached one side of the primary spark gap to an elevated wire (in effect, an antenna) and the other to Earth, with a similar arrangement for the secondary gap at the receiving point. The distance between transmitter and receiver was gradually increased first to 300 yards (275 m), then to 2 miles (3 km), then across the English Channel. Finally, in 1901, Marconi bridged the Atlantic when the letter *s* in Morse code traveled from Poldhu, Cornwall, to St. John's, Newfoundland, a distance of nearly 2,000 miles (3,200 km). For this distance, Marconi replaced the secondary-spark detector with a device known as a coherer, which had been invented by the French electrical engineer Edouard Branly in 1890. Branly's detector consisted of a tube filled with iron filings that coalesced, or "cohered," when a radio-frequency voltage was applied to the ends of the tube. The cohesion of the iron filings allowed the passage of current from an auxiliary power supply to operate a relay that reproduced the Morse signals. The coherer had to be regularly tapped to separate the filings and prepare them to react to the next radio-frequency signal.

THE FLEMING DIODE AND DE FOREST AUDION

The next major event was the discovery that an electrode operating at a positive voltage inside the evacuated envelope of a heated filament lamp would carry a current. Thomas Edison had noted that the bulb of such a lamp blackened near the positive electrode, but it was Sir John Ambrose Fleming, professor of electrical engineering at Imperial College, London, who explored the phenomenon and in 1904 discovered the one-directional current effect between a positively biased electrode, which he called the anode, and the heated filament; the electrons flowed from filament to anode only. Fleming called the device a diode because it contained two electrodes, the anode and the heated filament. He noted that when an alternating current was applied to the diode, only the positive halves of the waves were passed—that is, the wave was rectified (changed from alternating to direct current). The diode could also be used to detect radio-frequency signals since it suppressed half the radio-frequency wave and produced a pulsed direct current corresponding to the on and off of the Morse code transmitted signals. Fleming's discovery was the first step to the amplifier tube that revolutionized radio communication in the early part of the 20th century.

Fleming failed to appreciate the possibilities he had opened up and it was the American inventor Lee De Forest who in 1906 conceived the idea of interposing an open-meshed grid between the heated filament and positively biased anode, or plate, to control the flow of electrons. De Forest called his invention an Audion. With it he could obtain a large voltage change at the plate for a small voltage change on the grid electrode. This was a discovery of major importance because it made it possible to amplify the radio-frequency signal picked up by the antenna before application to the receiver detector; thus, much weaker signals could be utilized than had previously been possible.

Transmission of Speech

The first commercial company to be incorporated for the manufacture of radio apparatus was the Wireless Telegraph and Signal Company, Ltd. (England) in July 1897 (later changed to Marconi's Wireless Telegraph Company, Ltd.); other countries soon showed an interest in the commercial exploitation of radio.

Among the major developments of the first two decades of the 20th century was De Forest's discovery in 1912 of the oscillating properties of his Audion tube, a discovery that led to the replacement of the spark transmitter by an electronic tube oscillator that could generate much purer radio waves of relatively stable frequency. By 1910, radio messages between land stations and ships had become commonplace, and in that year the first air-to-ground radio contact was established from an aircraft. A landmark transmission came in 1918, when a radiotelegraph message from the Marconi long-wave station at Caernarvon, in Wales, was received in Australia, over a distance of 11,000 miles (17,700 km).

Though early experiments had shown that speech could be transmitted by radio, the first significant demonstration was not made until 1915 when the American

Telephone & Telegraph Company (AT&T) successfully transmitted speech signals from west to east across the Atlantic between Arlington, Virginia, and Paris. A year later, a radiotelephone message was conveyed to an aircraft flying near Brooklands (England) airfield. In 1919 a Marconi engineer spoke across the Atlantic in the reverse direction from Ballybunion, Ireland, to the United States.

From 1920 onward radio made phenomenal progress through research activities in Europe, North America, and Asia. The invention of the electron tube and later the transistor (1948) made possible remarkable developments.

TELEVISION

The dream of seeing distant places is as old as the human imagination. Priests in ancient Greece studied the entrails of birds, trying to see in them what the birds had seen when they flew over the horizon. They believed that their gods, sitting in comfort on Mount Olympus, were gifted with the ability to watch human activity all over the world. And the opening scene of William Shakespeare's play *Henry IV, Part 1* introduces the character Rumour, upon whom the other characters rely for news of what is happening in the far corners of England.

Mechanical Systems

For ages it remained a dream, and then television came along, beginning with an accidental discovery. In 1872, while investigating materials for use in the transatlantic cable, English telegraph worker Joseph May realized that a selenium wire was varying in its electrical conductivity. Further investigation showed that the change occurred when a beam of sunlight fell on the wire, which by chance had been placed on a table near the window. Although its importance was not realized at the time, this happenstance provided the basis for changing light into an electric signal.

In 1880 a French engineer named Maurice LeBlanc published an article in the journal *La Lumière électrique* that formed the basis of all subsequent television. LeBlanc proposed a scanning mechanism that would take advantage of the retina's temporary but finite retainment of a visual image. He envisaged a photoelectric cell that would look upon only one portion at a time of the picture to be transmitted. Starting at the upper left corner of the picture, the cell would proceed to the right-hand side and then jump back to the left-hand side, only one line lower. It would continue in this way, transmitting information on how much light was seen at each portion, until the entire picture was scanned, in a manner similar to the eye reading a page of text. A receiver would be synchronized with the transmitter, reconstructing the original image line by line.

The concept of scanning, which established the possibility of using only a single wire or channel for transmission of an entire image, became and remains to this day the basis of all television. LeBlanc, however, was never able to construct a working machine. Nor was the man who took television to the next stage: Paul Nipkow, a German engineer who invented the scanning disk. Nipkow's 1884 patent for an *Elektrisches Teleskop* was based on a simple rotating disk perforated with an inward-spiraling sequence of holes. It would be placed so that it blocked reflected light from the subject. As the disk rotated, the outermost hole would move across the scene, letting through light from the first "line" of the picture. The next hole would do the same thing slightly lower, and so on. One complete revolution of the disk would provide a complete picture, or "scan," of the subject.

This concept was eventually used by John Logie Baird in Britain and Charles Francis Jenkins in the United States to build the world's first successful televisions. The question of priority depends on one's definition of television. In 1922 Jenkins sent a still picture by radio waves, but the first true television success, the transmission of a live human face, was achieved by Baird in 1925. (The word *television* itself had been coined by a Frenchman, Constantin Perskyi, at the 1900 Paris Exhibition.)

The efforts of Jenkins and Baird were generally greeted with ridicule or apathy. As far back as 1880 an article in the British journal *Nature* had speculated that television was possible but not worthwhile: the cost of building a system would not be repaid, for there was no way to make money out of it. A later article in *Scientific American* thought there might be some uses for television, but entertainment was not one of them. Most people thought the concept was lunacy.

Nevertheless, the work went on and began to produce results and competitors. In 1927 the American Telephone and Telegraph Company (AT&T) gave a public demonstration of the new technology, and by 1928 the General Electric Company (GE) had begun regular television broadcasts. GE used a system designed by Ernst F. W. Alexanderson that offered "the amateur, provided with such receivers as he may design or acquire, an opportunity to pick up the signals," which were generally of smoke rising from a chimney or other such subjects. That same year Jenkins began to sell television kits by mail and established his own television station, showing cartoon pantomime programs. In 1929 Baird convinced the British Broadcasting Corporation (BBC) to allow him to produce half-hour shows at midnight three times a week. The following years saw the first "television boom," with thousands of viewers buying or constructing primitive sets to watch primitive programs.

Not everyone was entranced. C. P. Scott, editor of the *Manchester Guardian*, warned: "Television? The word is half Greek and half Latin. No good will come of it." More important, the lure of a new technology soon paled. The pictures, formed of only 30 lines repeating approximately 12 times per second, flickered badly on dim receiver screens only a few inches high. Programs were simple, repetitive, and ultimately boring. Nevertheless, even while the boom collapsed a competing development was taking place in the realm of the electron.

Electronic Systems

The final, insurmountable problems with any form of mechanical scanning were the limited number of scans per second, which produced a flickering image, and the relatively large size of each hole in the disk, which resulted in poor resolution. In 1908 a Scottish electrical engineer, A. A. Campbell Swinton, wrote that the problems "can probably be solved by the employment of two beams of kathode rays" instead of spinning disks. Cathode rays are beams of electrons generated in a vacuum tube. Steered by magnetic fields or electric fields, Swinton argued, they could "paint" a fleeting picture on the glass screen of a tube coated on the inside with a phosphorescent material. Because the rays move at nearly the speed of light, they would avoid the flickering problem, and their tiny size would allow excellent resolution. Swinton never built a set (for, as he said, the possible financial reward would not be enough to make it worthwhile), but unknown to him such work had already begun in Russia. In 1907 Boris Rosing, a lecturer at the St. Petersburg Institute of Technology, put together equipment consisting of a mechanical scanner and a cathode-ray-tube receiver. There is no record of

Rosing actually demonstrating a working television, but he had an interested student named Vladimir Kosma Zworykin, who soon immigrated to America.

In 1923, while working for the Westinghouse Electric Company in Pittsburgh, Pennsylvania, Zworykin filed a patent application for an all-electronic television system, although he was as yet unable to build and demonstrate it. In 1929 he convinced David Sarnoff, vice president and general manager of Westinghouse's parent company, the Radio Corporation of America (RCA), to support his research by predicting that in two years, with $100,000 of funding, he could produce a workable electronic television system. Meanwhile, the first demonstration of a primitive electronic system had been made in San Francisco in 1927 by American inventor Philo Farnsworth, a young man with only a high-school education. Farnsworth had garnered research funds by convincing his investors that he could market an economically viable television system in six months for an investment of only $5,000. In actuality, it took the efforts of both men and more than $50 million before anyone made a profit.

With his first $100,000 of RCA research money, Zworykin developed a workable cathode-ray receiver that he called the Kinescope. At the same time, Farnsworth was perfecting his Image Dissector camera tube. In 1930 Zworykin visited Farnsworth's laboratory and was given a demonstration of the Image Dissector. At that point a healthy cooperation might have arisen between the two pioneers, but competition, spurred by the vision of corporate profits, kept them apart. Sarnoff offered Farnsworth $100,000 for his patents but was summarily turned down. Farnsworth instead accepted an offer to join RCA's rival Philco, but he soon left to set up his own firm. Then in 1931 Zworykin's RCA team, after learning much from the study of Farnsworth's Image

PHILO FARNSWORTH

Philo Taylor Farnsworth was born into a family of Mormon farmers on Aug. 19, 1906, in Beaver, Utah. While still in high school, Farnsworth conceived the basic requirements for television, sketching out the basic design of an electronic image scanner for the benefit of his chemistry teacher. After two years at Brigham Young University, Provo, Utah, he began research into the process of picture transmission. In 1926 he cofounded Crocker Research Laboratories, which was reorganized as Farnsworth Television, Inc. (1929), and later as Farnsworth Radio and Television Corporation (1938).

In 1927, in his own small laboratory in San Francisco, Farnsworth successfully transmitted an image (a dollar sign) composed of 60 horizontal lines and submitted his first television patent, for a camera tube he called the Image Dissector. The patent was granted in 1930. Farnsworth subsequently invented numerous other devices, including amplifiers, cathode-ray tubes, vacuum tubes, electrical scanners, electron multipliers, and photoelectric materials. He held some 165 patents at the time of his death in Salt Lake City, Utah, on March 11, 1971.

Dissector, came up with the Iconoscope camera tube, and with it they finally had a working electronic system.

In England the Gramophone Company, Ltd., and the London branch of the Columbia Phonograph Company joined in 1931 to form Electric and Musical Industries, Ltd. (EMI). Through the Gramophone Company's ties with

RCA-Victor, EMI was privy to Zworykin's research, and soon a team under Isaac Shoenberg produced a complete and practical electronic system, reproducing moving images on a cathode-ray tube at 405 lines per picture and 25 pictures per second. Baird excoriated this intrusion of a "non-English" system, but he reluctantly began research on his own system of 240-line pictures by inviting a collaboration with Farnsworth. On Nov. 2, 1936, the BBC instituted an electronic TV competition between Baird and EMI, broadcasting the two systems from the Alexandra Palace (called for the occasion the "world's first, public, regular, high-definition television station"). Several weeks later a fire destroyed Baird's laboratories. EMI was declared the victor and went on to monopolize the BBC's interest. Baird never really recovered; he died several years later, nearly forgotten and destitute.

By 1932 the conflict between RCA and Farnsworth had moved to the courts, both sides claiming the invention of electronic television. Years later the suit was finally ruled in favour of Farnsworth, and in 1939 RCA signed a patent-licensing agreement with Farnsworth Television and Radio, Inc. This was the first time RCA ever agreed to pay royalties to another company. But RCA, with its great production capability and estimable public-relations budget, was able to take the lion's share of the credit for creating television. At the 1939 World's Fair in New York City, Sarnoff inaugurated America's first regular electronic broadcasting, and 10 days later, at the official opening ceremonies, Franklin D. Roosevelt became the first U.S. president to be televised.

THE TRANSISTOR

The transistor is a semiconductor device for amplifying, controlling, and generating electrical signals. It was invented in 1947–48 by three American physicists, John Bardeen, Walter H. Brattain, and William B. Shockley, at the American Telephone and Telegraph Company's Bell Laboratories. The transistor proved to be a viable alternative to the electron tube and, by the late 1950s, supplanted the latter in many applications. Its small size, low heat generation, high reliability, and low power consumption made possible a breakthrough in the miniaturization of complex circuitry. During the 1960s and '70s, transistors were incorporated into integrated circuits, in which a multitude of components (e.g., diodes, resistors, and capacitors) are formed on a single "chip" of semiconductor material. Deeply embedded in almost everything electronic, transistors have become the nerve cells of the Information Age.

Motivation and Early Radar Research

Electron tubes are bulky and fragile, and consume large amounts of power to heat their cathode filaments and generate streams of electrons. They also often burn out after several thousand hours of operation. Electromechanical switches, or relays, are slow and can become stuck in the on or off position. For applications requiring thousands of tubes or switches, such as the nationwide telephone systems developing around the world in the 1940s and the first electronic digital computers, this meant constant vigilance was needed to minimize the inevitable breakdowns.

An alternative was found in semiconductors, materials such as silicon or germanium whose electrical conductivity lies midway between that of insulators such as glass and conductors such as aluminum. The conductive properties of semiconductors can be controlled by "doping" them with select impurities, and a few visionaries had seen the potential of such devices for telecommunications and computers. However, it was military funding for radar development in the 1940s that opened the door to their realization. The "superheterodyne" electronic circuits used to detect radar waves required a diode rectifier—a device that allows current to flow in just one direction—that could operate successfully at ultrahigh frequencies over 1 gigahertz. Electron tubes did not suffice, and solid-state diodes based on existing copper-oxide semiconductors were also much too slow for this purpose.

Crystal rectifiers based on silicon and germanium came to the rescue. In these devices a tungsten wire was jabbed into the surface of the semiconductor material, which was doped with tiny amounts of impurities, such as boron or phosphorus. The impurity atoms assumed positions in the material's crystal lattice, displacing silicon (or germanium) atoms and thereby generating tiny populations of charge carriers (such as electrons) capable of conducting usable electrical current. Depending on the nature of the charge carriers and the applied voltage, a current could flow from the wire into the surface or vice versa, but not in both directions. Thus, these devices served as the much-needed rectifiers operating at the gigahertz frequencies required for detecting rebounding microwave radiation in military radar systems. By the end of World War II, millions of crystal rectifiers were being produced annually by such American manufacturers as Sylvania and Western Electric.

Innovation at Bell Labs

Executives at Bell Labs had recognized that semiconductors might lead to solid-state alternatives to the electron-tube amplifiers and electromechanical switches employed throughout the nationwide Bell telephone system. In 1936 the new director of research at Bell Labs, Mervin Kelly, began recruiting solid-state physicists. Among his first recruits was William B. Shockley, who proposed a few amplifier designs based on copper-oxide semiconductor materials then used to make diodes. With the help of Walter H. Brattain, an experimental physicist already working at Bell Labs, he even tried to fabricate a prototype device in 1939, but it failed completely. Semiconductor theory could not yet explain exactly what was happening to electrons inside these devices, especially at the interface between copper and its oxide. Compounding the difficulty of any theoretical understanding was the problem of controlling the exact composition of these early semiconductor materials, which were binary combinations of different chemical elements (such as copper and oxygen).

With the close of World War II, Kelly reorganized Bell Labs and created a new solid-state research group headed by Shockley. The postwar search for a solid-state amplifier began in April 1945 with Shockley's suggestion that silicon and germanium semiconductors could be used to make a field-effect amplifier (*see* integrated circuit: Field-effect transistors). He reasoned that an electric field from a third electrode could increase the conductivity of a sliver of semiconductor material just beneath it and thereby allow usable current to flow through the sliver. But attempts to fabricate such a device by Brattain and others in Shockley's group again failed. The following March, John Bardeen, a theoretical physicist whom Shockley had hired for his group, offered a possible explanation. Perhaps electrons drawn to the semiconductor surface by the electric field were blocking the penetration of this field into the bulk material, thereby preventing it from influencing the conductivity.

Bardeen's conjecture spurred a basic research program at Bell Labs into the behaviour of these "surface-state" electrons. While studying this phenomenon in November

1947, Brattain stumbled upon a way to neutralize their blocking effect and permit the applied field to penetrate deep into the semiconductor material. Working closely together over the next month, Bardeen and Brattain invented the first successful semiconductor amplifier, called the point-contact transistor, on Dec. 16, 1947. Similar to the World War II crystal rectifiers, this weird-looking device had not one but two closely spaced metal wires jabbing into the surface of a semiconductor—in this case, germanium. The input signal on one of these wires (the emitter) boosted the conductivity of the germanium beneath both of them, thus modulating the output signal on the other wire (the collector). Observers present at a demonstration of this device the following week could hear amplified voices in the earphones that it powered. Shockley later called this invention a "magnificent Christmas present" for the farsighted company, which had supported the research program that made this breakthrough.

Not to be outdone by members of his own group, Shockley conceived yet another way to fabricate a semiconductor amplifier the very next month, on Jan. 23, 1948. His junction transistor was basically a three-layer sandwich of germanium or silicon in which the adjacent layers would be doped with different impurities to induce distinct electrical characteristics. An input signal entering the middle layer—the "meat" of the semiconductor sandwich—determined how much current flowed from one end of the device to the other under the influence of an applied voltage. Shockley's device is often called the bipolar junction transistor because its operation requires that the negatively charged electrons and their positively charged counterparts (the holes corresponding to an absence of electrons in the crystal lattice) coexist briefly in the presence of one another.

The name *transistor*, a combination of *transfer* and *resistor*, was coined for these devices in May 1948 by Bell Labs electrical engineer John Robinson Pierce, who was also a science-fiction author in his spare time. A month later Bell Labs announced the revolutionary invention in a press conference held at its New York City headquarters, heralding Bardeen, Brattain, and Shockley as the three coinventors of the transistor. The three were eventually awarded the Nobel Prize for Physics for their invention.

Although the point-contact transistor was the first transistor invented, it faced a difficult gestation period and was eventually used only in a switch made for the Bell telephone system. Manufacturing them reliably and with uniform operating characteristics proved a daunting problem, largely because of hard-to-control variations in the metal-to-semiconductor point contacts.

Shockley had foreseen these difficulties in the process of conceiving the junction transistor, which he figured would be much easier to manufacture. But it still required more than three years, until mid-1951, to resolve its own development problems. Bell Labs scientists, engineers, and technicians first had to find ways to make ultrapure germanium and silicon, form large crystals of these elements, dope them with narrow layers of the required impurities, and attach delicate wires to these layers to serve as electrodes. In July 1951 Bell Labs announced the successful invention and development of the junction transistor, this time with only Shockley in the spotlight.

THE COMMUNICATIONS SATELLITE

The idea of radio transmission through space is at least as old as the space novel *Ralph 124C41+* (1911), by the American science fiction pioneer Hugo Gernsback. Yet the idea of a radio repeater located in space was slow to develop. In

1945 the British author and scientist Arthur C. Clarke proposed the use of geostationary satellites for station-to-station and broadcast radio communication. Clarke assumed that these spacecraft would need to take the form of manned space stations with living quarters for crew that would be built in space of materials flown up by rockets and provided with receiving and transmitting equipment and directional antennas. Clarke suggested the use of solar power, either a steam engine operated by solar heat or photoelectric devices.

In a paper published in April 1955, the American engineer and scientist John Robinson Pierce analyzed various concepts for unmanned communications satellites. These included passive devices, such as metallized balloons, plane reflectors, and corner reflectors, that would merely reflect back to Earth part of the energy directed to them. Active satellites, incorporating radio receivers and transmitters, were also considered. Pierce discussed satellites at synchronous altitudes, satellites at lower altitudes, and the use of Earth's gravity to control the attitude or orientation of a satellite.

The first satellite to relay messages between Earth stations was the U.S. government's Project SCORE, launched Dec. 18, 1958. Circling Earth in a low elliptical orbit, it functioned for 13 days until its batteries ran down. One of the best-known early satellites was Echo 1, a balloon made of Mylar plastic coated with a thin layer of aluminum, which was launched Aug. 12, 1960. Successful communications tests carried out by reflecting radio signals from Echo 1's surface encouraged further experimentation. Telstar 1, launched July 10, 1962, was an active satellite and was the first to transmit live television signals and telephone conversations across the Atlantic Ocean. Syncom 2, launched July 26, 1963, was the first geostationary communications satellite, and Syncom 3, launched

Communication

The American-built Telstar 1 communications satellite, launched July 10, 1962, relayed the first transatlantic television signals. NASA

Aug. 19, 1964, relayed the first sustained transpacific television picture.

Experimental programs such as those described above represented the conjunction of a number of technological advances that were necessary for the era of satellite communication to begin. These included the development of reliable launch vehicles, solid-state electronic devices, spin stabilization for attitude control, efficient solar cells for power generation, and Earth-to-satellite telemetering and control techniques. In addition, the development of the low-noise maser and the traveling-wave tube amplifier was necessary for satellites to capture and amplify the weak uplink signals for retransmission to Earth. Intelsat 1 (also known as Early Bird), the first commercial communications satellite, was launched April 6, 1965. It provided high-bandwidth telecommunications service between the United States and Europe as a supplement to existing transatlantic cable and shortwave radio links. Intelsat 1 carried 240 voice circuits or one television channel. The Intelsat 2 series of satellites (launched 1967) together offered full coverage of the Atlantic and Pacific regions, and each satellite of the Intelsat 3 series (1968–70) provided more than 1,500 voice circuits or four television channels. The Intelsat 4 satellites (1971–75) each carried 6,000 voice circuits or 12 television channels. In contrast to these early series, the Intelsat 9 satellites launched in the first years of the 21st century each could handle 600,000 circuits or 600 television channels.

THE PERSONAL COMPUTER

The personal computer, or PC, is a digital computer designed for use by only one person at a time. A typical personal computer assemblage consists of a central processing unit (CPU), which contains the computer's arithmetic, logic, and control circuitry on an integrated circuit. There are two types of computer memory: main memory, such as digital random-access memory (RAM), and auxiliary memory, such as magnetic hard disks and special optical compact discs, or read-only memory (ROM) discs (CD-ROMs and DVD-ROMs). Personal computers also have various input/output devices, including a display screen, keyboard and mouse, modem, and printer.

Computers small and inexpensive enough to be purchased by individuals for use in their homes first became feasible in the 1970s, when large-scale integration made it possible to construct a sufficiently powerful microprocessor on a single semiconductor chip. A small firm named MITS made the first personal computer, the Altair. This computer, which used Intel Corporation's 8080 microprocessor, was developed in 1974. Though the Altair was popular among computer hobbyists, its commercial appeal was limited, since purchasers had to assemble the machine from a kit.

The personal computer industry truly began in 1977, when Apple Computer, Inc. (now Apple, Inc.), founded by Steven P. Jobs and Stephen G. Wozniak, introduced the Apple II, one of the first pre-assembled, mass-produced personal computers. Radio Shack and Commodore Business Machines also introduced personal computers that year. These machines used 8-bit microprocessors (which process information in groups of 8 bits, or binary digits, at a time) and possessed rather limited memory capacity—i.e., the ability to address a given quantity of data held in memory storage. But because personal computers were much less expensive than mainframe computers (the bigger computers typically deployed by large business, industry, and government organizations), they could be purchased by individuals, small and medium-sized businesses, and primary and secondary schools. The Apple II received a great boost in popularity when it became the host machine for VisiCalc, the first electronic spreadsheet (computerized accounting program). Other types of application software soon developed for personal computers.

IBM Corporation, the world's dominant computer maker, did not enter the new market until 1981, when it introduced the IBM Personal Computer, or IBM PC. The IBM PC was only slightly faster than rival machines, but it had about 10 times their memory capacity, and it was backed by IBM's large sales organization. The IBM PC was also the host machine for 1-2-3, an extremely popular spreadsheet introduced by the Lotus Development Corporation in 1982. The IBM PC became the world's most popular personal computer, and both its microprocessor, the Intel 8088, and its operating system, which was adapted from Microsoft Corporation's MS-DOS system, became industry standards. Rival machines that used Intel microprocessors and MS-DOS became known as "IBM compatibles" if they tried to compete with IBM on the basis of additional computing power or memory and "IBM clones" if they competed simply on the basis of low price.

In 1983 Apple introduced Lisa, a personal computer with a graphical user interface (GUI) to perform routine operations. A GUI is a display format that allows the user to select commands, call up files, start programs, and do other routine tasks by using a device called a mouse to point to pictorial symbols (icons) or lists of menu choices on the screen. This type of format had certain advantages over interfaces in which the user typed text- or character-based commands on a keyboard to perform routine tasks.

A GUI's windows, pull-down menus, dialog boxes, and other controlling mechanisms could be used in new programs and applications in a standardized way so that common tasks were always performed in the same manner. The Lisa's GUI became the basis of Apple's Macintosh personal computer, which was introduced in 1984 and proved extremely successful. The Macintosh was particularly useful for desktop publishing because it could lay out text and graphics on the display screen as they would appear on the printed page.

The Macintosh's graphical interface style was widely adapted by other manufacturers of personal computers and PC software. In 1985 the Microsoft Corporation introduced Microsoft Windows, a graphical user interface that gave MS-DOS-based computers many of the same capabilities of the Macintosh. Windows became the dominant operating environment for personal computers.

These advances in software and operating systems were matched by the development of microprocessors containing ever-greater numbers of circuits, with resulting increases in the processing speed and power of personal computers. The Intel 80386 32-bit microprocessor (introduced 1985) gave the Compaq Computer Corporation's Compaq 386 (introduced 1986) and IBM's PS/2 family of computers (introduced 1987) greater speed and memory capacity. Apple's Mac II computer family made equivalent advances with microprocessors made by Motorola, Inc. The memory capacity of personal computers had increased from 64 kilobytes (64,000 characters) in the late 1970s to 100 megabytes (100 million characters) by the early '90s to several gigabytes (billions of characters) by the early 2000s.

By 1990 some personal computers had become small enough to be completely portable. They included laptop

THE APPLE II

Apple, Inc., had its genesis in the lifelong dream of Stephen G. Wozniak to build his own computer—a dream that was made suddenly feasible with the arrival in 1975 of the first commercially successful microcomputer, the Altair 8800, which came as a kit and used the recently invented microprocessor chip. Encouraged by his friends at the Homebrew Computer Club, a San Francisco Bay area group centred around the Altair, Wozniak quickly came up with a plan for his own microcomputer. In 1976, when the Hewlett-Packard Company, where Wozniak was an engineering intern, expressed no interest in his design, Wozniak, then 26 years old, together with a former high-school classmate, 21-year-old Steven P. Jobs, moved production operations to the Jobs family garage—and the Silicon Valley garage start-up company legend was born. Jobs and Wozniak named their company Apple. For working capital, Jobs sold his Volkswagen minibus and Wozniak his programmable calculator. Their first model was simply a working circuit board, but at Jobs's insistence the 1977 version was a stand-alone machine in a custom-molded plastic case, in contrast to the forbidding steel boxes of other early machines. This Apple II also offered a colour display and other features that made Wozniak's creation the first microcomputer that appealed to the average person.

computers, also known as notebook computers, which were about the size of a notebook, and less-powerful pocket-sized computers, known as personal digital assistants (PDAs). At the high end of the PC market, multimedia personal computers equipped with DVD players and digital sound systems allowed users to handle animated images and sound (in addition to text and still images) that were stored on high-capacity DVD-ROMs. Personal computers were increasingly interconnected with each other and with larger computers in networks for the purpose of gathering, sending, and sharing information electronically. The uses of personal computers continued to multiply as the machines became more powerful and their application software proliferated.

By 2000 more than 50 percent of all households in the United States owned a personal computer, and this penetration increased dramatically over the next few years as people in the United States (and around the world) purchased PCs to access the world of information available through the Internet.

The Internet

The Internet grew out of funding by the U.S. Advanced Research Projects Agency (ARPA), later renamed the Defense Advanced Research Projects Agency (DARPA), to develop a communication system among government and academic computer-research laboratories. The first network component, ARPANET, became operational in October 1969. ARPANET soon became a critical piece of infrastructure for the computer science research community in the United States. Tools and applications—such as the simple mail transfer protocol (SMTP, commonly referred to as e-mail), for sending short messages, and the file transfer protocol (FTP), for longer transmissions—quickly emerged. In order to achieve cost-effective interactive communications between computers, which typically communicate in short bursts of data, ARPANET employed the new technology of packet switching. Packet

VIRTUAL COMMUNITIES

The first use of the phrase "virtual community" appeared in a 1987 article in *The Whole Earth Review* written by Howard Rheingold, who used the term to describe any group of people who may or may not meet one another face-to-face but who exchange words and ideas through the mediation of computer bulletin board systems (BBSs) and other digital networks. In *The Virtual Community* (1993), Rheingold expanded on his article to offer the following definition:

> "Virtual communities are social aggregations that emerge from the Net when enough people carry on those public discussions long enough, with sufficient human feeling, to form webs of personal relationships in cyberspace."

Today, with several billion mobile telephones now in existence and a growing trend toward building Internet connections into mobile devices, it is not unlikely that a significant portion of the human population will conduct some of their social affairs by means of computer networks. In the 21st century, people meet, play, conduct discourse, socialize, do business, and organize collective action through instant messages, blogs (including videoblogs), RSS feeds (a format for subscribing to and receiving regularly updated content from Web sites), wikis, social network services such as MySpace and Facebook, photo- and media-sharing communities such as Flickr, massive multiplayer online games such as *Lineage* and *World of Warcraft*, and immersive virtual worlds such as Second Life. Virtual communities and social media have coevolved as emerging technologies have afforded new kinds of interaction and as different groups of people have appropriated media for new purposes.

As the early digital enthusiasts, builders, and researchers have been joined by a more representative sample of the world's population, a broader and not always wholesome representation of human behaviour has manifested itself online. Life online in the 21st century has enabled terrorists and various cybercriminals to make use of the same many-to-many digital networks that enable support groups for disease victims and caregivers, disaster relief action, distance learning, and community-building efforts. With so many young people spending so much of their time online, many parents and "real world" community leaders have expressed concerns about the possible effects of overindulging in such virtual social lives. In addition, in an environment where anyone can publish anything or make any claim online, the need to include an understanding of social media in education has given rise to advocates for "participatory pedagogy."

Students of online social behaviour have noted a shift from "group-centric" characterizations of online socializing to a perspective that takes into account "networked individualism." To quote Canadian sociologist Barry Wellman:

> "Although people often view the world in terms of groups, they function in networks. In networked societies: boundaries are permeable, interactions are with diverse others, connections switch between multiple networks, and hierarchies can be flatter and recursive.... Most people operate in multiple, thinly connected, partial communities as they deal with networks of kin, neighbours, friends, workmates, and organizational ties. Rather than fitting into the same group as those around them, each person has his/her own 'personal community.'"

> It is likely that community-centred forms of online communication will continue to flourish. At the same time, it is also likely that the prevalence of individual-centred social network services and the proliferation of personal communication devices will feed the evolution of "networked individualism." Cyberculture studies, necessarily an interdisciplinary pursuit, is likely to continue to grow as more human socialization is mediated by digital networks.

switching takes large messages (or chunks of computer data) and breaks them into smaller, manageable pieces (known as packets) that can travel independently over any available circuit to the target destination, where the pieces are reassembled. Thus, unlike traditional voice communications, packet switching does not require a single dedicated circuit between each pair of users.

With only 15 nongovernment (university) sites included in ARPANET, the U.S. National Science Foundation decided to fund the construction and initial maintenance cost of a supplementary network, the Computer Science Network (CSNET). Built in 1980, CSNET was made available on a subscription basis to a wide array of academic, government, and industry research labs. As the 1980s wore on, further networks were added. In North America there were (among others): BITNET (Because It's Time Network) from IBM, UUCP (UNIX-to-UNIX Copy Protocol) from Bell Telephone, USENET (initially a connection between Duke University, Durham, North Carolina, and the University of North Carolina and still the home system for the Internet's many newsgroups), NSFNET (a high-speed National Science Foundation network connecting supercomputers), and CDNet (in Canada). In Europe several small academic networks were linked to the growing North American network.

In order to connect these various networks, DARPA established a program to investigate the interconnection of "heterogeneous networks." This program, called Internetting, was based on the newly introduced concept of open architecture networking, in which networks with defined standard interfaces would be interconnected by "gateways." A working demonstration of the concept was planned. In order for the concept to work, however, a new protocol had to be designed and developed; indeed, a system architecture was also required. In 1974 Vinton Cerf, then at Stanford University in California, and Robert Kahn, then at DARPA, collaborated on a paper that first described such a protocol and system architecture—namely, the transmission control protocol (TCP), which enabled different types of machines on networks all over the world to route and assemble data packets. TCP, which originally included the Internet protocol (IP), a global addressing mechanism that allowed routers to get data packets to their ultimate destination, formed the TCP/IP standard, which was adopted by the U.S. Department of Defense in 1980. By the early 1980s the "open architecture" of the TCP/IP approach was adopted and endorsed by many other researchers and eventually by technologists and businessmen around the world.

The rapid commercialization of the Internet was the result of several factors. One important factor was the introduction of the personal computer and the workstation in the early 1980s—a development that in turn was fueled by unprecedented progress in integrated circuit technology and an attendant rapid decline in computer prices. Another factor, which took on increasing importance, was the emergence of ethernet and other "local area networks" to link personal computers. But what it took to turn a network of computers into something more was the idea of the hyperlink: computer code inside a document that would cause related documents to be fetched and displayed. The concept of hyperlinking was anticipated from the early to the middle decades of the 20th century—in Belgium by Paul Otlet and in the United States by Ted Nelson, Vannevar Bush, and, to some extent, Douglas Engelbart. Their yearning for some kind of system to link knowledge together, though, did not materialize until 1990, when Tim Berners-Lee of England and others at CERN (European Organization for Nuclear Research) developed a protocol based on hypertext to make information distribution easier. In 1991 this culminated in the creation of the World Wide Web and its system of links among user-created pages. A team of programmers at the U.S. National Center for Supercomputing Applications, Urbana, Ill., developed a program, called a browser, that ran on most types of computers. Through its "point-and-click" interface, the browser simplified access, retrieval, and display of files through the World Wide Web. The browser was called Mosaic. In 1994 Netscape Communications Corporation (originally called Mosaic Communications Corporation) was formed to further develop the Mosaic browser and server software for commercial use.

Netscape was an enormous success. The Web grew exponentially, doubling the number of users and the number of sites every few months. Uniform resource locators (URLs) became part of daily life, and the use of electronic mail (e-mail) became commonplace. Increasingly business took advantage of the Internet and adopted new forms of buying and selling in cyberspace, a term popularized in the early 1980s by science fiction author William Gibson. With Netscape so successful, Microsoft and other firms developed alternative Web browsers.

Thus the Internet, originally created as a closed network for researchers, was suddenly a new public medium for information. It became the home of virtual shopping malls, bookstores, stockbrokers, newspapers, and entertainment. Schools were "getting connected" to the Internet, and children were learning to do research in novel ways. The combination of the Internet, e-mail, and small and affordable computing and communication devices began to change many aspects of society.

Chapter 2: Transportation

As with communication, our means of moving ourselves and our goods from place to place have grown faster and more comprehensive with every passing generation of technology. We have progressed from a reliance on human and animal muscle to the harnessing of combustion to power automobiles, aircraft, and rockets. All of the inventions in this progression, from wheeled chariots to rocket launchers and jetliners, have extended our reach ever farther.

THE WHEELED CHARIOT

Archaeological evidence suggests that the first vehicles were heavy two- or four-wheeled chariots that were pulled by oxen. These vehicles seem to have been used first by the Sumerians of Mesopotamia about 3000 BCE. These chariots were of little use in ordinary life, but they probably were rather quickly adopted for military purposes and, by extension, to a symbolic role among the gods.

Areas in the Middle East and along the Mediterranean Sea (Mesopotamia, Syria, and Turkey), already deforested in classical times, were the typical locale for chariot warfare, which might involve up to several thousand charioteers in important battles. Early chariots featured a dashboard where the rider secured a handhold to keep himself from being pitched off the springless vehicle. The dashboard became increasingly horizontal, perhaps anticipating a four-wheel vehicle closer in form to a wagon. The spoked wheel allowed the war chariot to be much lighter and thereby capable of being drawn by mules rather than oxen and ultimately by the much faster horse. The spoked wheel was introduced in the 2nd millennium BCE, after which time the horse chariot was developed. In that undertaking, however, a new problem arose. Earlier chariots had been yoked to the ox at the withers, but such a yoke did not rest securely on a horse. Instead a breastband was introduced, but it had a tendency to choke the horse and greatly reduce the power transferred to the chariot. Only many centuries later, in the Middle Ages, was harnessing improved to the point where the full power of the horse was made available to gain either greater speed or a stronger pull.

The development of earlier chariots was most strongly associated with Ur and southern Mesopotamia, as well as with the nomadic steppe dwellers, but further advances were made by the Hittites and Assyrians. As bronze spokes were more commonly used (to enhance speed), the necessity for better roads increased. For military purposes, road builders used bronze pickaxes to cut into the side of a steep hill slope in the 2nd millennium BCE in Assyria. By 600 BCE Assyrian military domination, which had relied upon swift chariots and understanding of tactics, was over. The Babylonians and the Urartians (living in what is now Armenia) developed new forms of chariots. Somewhat later the Persians created a four-poled chariot drawn by eight horses that became a weapon of terror; it incorporated a mower with blades attached to the rotating wheels that literally slashed to shreds the opposing infantry.

No doubt there were civilian uses of these vehicles, but relatively little evidence remains. It was the military and religious (to transport statues of gods) uses of chariots that assured the creation of pictorial representations. It is known, however, that Alexander the Great employed a considerable number of freight wagons on his campaigns.

The chariot was introduced in Egypt from Syria during the period 1670–1570 BCE. Quickly the pharaohs adopted the vehicle, first to defend themselves against similarly equipped forces but soon thereafter as "showpieces" displaying physical evidence of Egypt's might. Once well established in Syria, Armenia, Anatolia, and Egypt, the war chariot spread farther into the Sahara, and finally across the Aegean into Greece and then into the Roman world. There it played a significant role in the conquest that ultimately shaped the Roman Empire, the continued success of which was firmly established on a carefully planned and assiduously executed system of roads and the vehicles that circulated on them.

SAILING SHIPS

The earliest historical evidence of boats is found in Egypt during the 4th millennium BCE. A culture nearly completely riparian, Egypt was narrowly aligned along the Nile, totally supported by it, and served by transport on its uninterruptedly navigable surface below the First Cataract (at modern-day Aswan). There are representations of Egyptian boats used to carry obelisks on the Nile from Upper Egypt that were as long as 300 feet (100 m), longer than any warship constructed in the era of wooden ships.

From Rowed Vessels to Sails

The Egyptian boats commonly featured sails as well as oars. Because they were confined to the Nile and depended on winds in a narrow channel, the ability to row was essential. This became true of most navigation when the Egyptians began to venture out onto the shallow waters of the Mediterranean and Red seas. Most early Nile boats had a single square sail as well as one level, or row, of oarsmen. Quickly, several levels came into use, as it was difficult to maneuver very elongated boats in the open sea. The later Roman two-level bireme and three-level trireme were most common, but sometimes more than a dozen banks of oars were used to propel the largest boats.

The basic functions of the warship and cargo ship determined their design. Because fighting ships required speed, adequate space for substantial numbers of fighting men, and the ability to maneuver at any time in any direction, long, narrow rowed ships became the standard for naval warfare. In contrast, because trading ships sought to carry as much tonnage of goods as possible with as small a crew as practicable, the trading vessel became as round a ship as could be easily navigated. The trading vessel required increased freeboard (height between the waterline and upper deck level), as the swell in the larger seas could fairly easily swamp the low-sided galleys propelled by oarsmen.

As was true of early wheeled vehicles, ship design also showed strong geographic orientation. In the conquest of

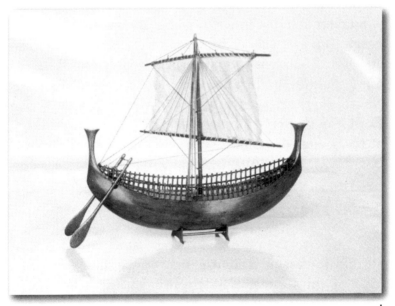

Model of a Phoenician ship, 13th century BCE, in the Museum of the Philadelphia Civic Center. Courtesy of the Museum of the Philadelphia Civic Center

Britain and in their encounter with the Batavian area in Holland, Romans became aware of the northern European boat. It was generally of clinker construction (that is, with a hull built of overlapping timbers) and identical at either end. In the Mediterranean, ship design favoured carvel-built (that is, built of planks joined along their lengths to form a smooth surface) vessels that differed at the bow and stern (the forward and rear ends, respectively). In the early centuries, both Mediterranean and northern boats were commonly rowed, but the cyclonic storms found year-round in the Baltic and North Sea latitudes encouraged the use of sails. Because the sailing techniques of these early centuries depended heavily on sailing with a following wind (i.e., from behind), the frequent shifts in wind direction in the north permitted, after only relatively short waits, navigation in most compass directions. In the persistent summer high-pressure systems of the Mediterranean, however, the long waits for a change of wind direction discouraged sailing. It was also more economical to carry goods by ship in the north. With a less absolute dependence on rowing, the double-ended clinker boat could be built with a greater freeboard than was possible in the rowed galleys of the Mediterranean. When European sailors began to look with increasing curiosity at the seemingly boundless Atlantic Ocean, greater freeboard made oceanic navigation more practicable.

Types of Sails

The move to the pure sailing ship came with small but steadily increasing technical innovations that more often allowed ships to sail with the wind behind them. Sails changed from a large square canvas suspended from a single yard (top spar), to complex arrangements intended to pivot on the mast depending on the direction and force of the wind. Instead of being driven solely by the wind direction, ships could "sail into the wind" to the extent that the course taken by a ship became the product of a resolution of forces (the actual wind direction and the objective course of the particular ship). Sails were devised to handle gentle breezes and to gain some mileage from them as well as from strong winds and to maintain some choice as to course while under their influence.

While the speed of a rowed ship was mainly determined by the number of oarsmen in the crew, in sailing ships the total spread of canvas in the sails was the main determinant of speed. Because winds are not fixed either as to direction or as to force, gaining the maximum effective propulsion from them requires complexly variable sails. There was one constant that characterized navigation by sail throughout its history—to gain speed it was necessary to increase the number of masts on the ship. Ships in both the Mediterranean and the north were single-masted until about 1400 CE and likely as well to be rigged for one basic type of sail. With experience square sails replaced the simple lateen sails that were the mainstay during the Middle Ages, particularly in the Mediterranean.

In the earlier centuries of sailing ships the dominant rig was the square sail, which features a canvas suspended on a boom, held aloft by the mast, and hung across the longitudinal axis of the ship. To utilize the shifting relationship between the desired course of the ship and the present wind direction, the square sail must be twisted on the mast to present an edge to the wind. This meant that most ships had to have clear decks amidships to permit the shifting of the sail and its boom; most of the deck space was thus monopolized by a single swinging sail. Large sails also required a sizable gang of men to raise and

THE CHINESE JUNK

During the period that the sailing ship was developing in the Mediterranean world, China, with its vast land areas and poor road communications, was turning to water for transportation. Starting with a dugout canoe, the Chinese joined two canoes with planking, forming a square punt, or raft. Next, the side, bow, and stern were built up with planking to form a large, flat-bottomed wooden box. The bow was sharpened with a wedge-shaped addition below the waterline. At the stern, instead of merely hanging a steering oar over one side as did the Western ships, Chinese shipbuilders contrived a watertight box, extending through the deck and bottom, that allowed the steering oar or rudder to be placed on the centreline, thus giving better control. The stern was built to a high, small platform at the stern deck, later called a castle in the West, so that, in a following sea, the ship would remain dry. Thus, in spite of what to Western eyes seemed an ungainly figure, the "Chinese junk" was an excellent hull for seaworthiness as well as for beaching in shoal (shallow) water. The principal advantage, however, not apparent from an external view, was great structural rigidity. In order to support the side and the bow planking, the Chinese used solid planked walls (bulkheads), running both longitudinally and transversely and dividing the ship into 12 or more compartments. This produced not only strength but also protection against damage.

In rigging the Chinese junk was far ahead of Western ships, with sails made of narrow panels, each tied to a sheet (line) at each end so that the force of the wind could be taken in many lines rather than on the mast alone. Also, the sail could be hauled about to permit the ship to sail somewhat into the wind. By the 15th century junks had developed into the largest, strongest, and most seaworthy ships in the world. Not until about the 19th century did Western ships catch up in performance.

lower the sail (and, when reef ports were introduced, to reef the sail—that is, to reduce its area by gathering up the sail at the reef points).

By 1200 the standard sailing ship in the Mediterranean was two-masted, with the foremast larger and hung with a sail new to ordinary navigation at sea. This was the lateen sail, earlier known to the Egyptians and sailors of the eastern Mediterranean. The lateen sail is triangular in shape and is fixed to a long yard mounted at its middle to the top of the mast. The combination of sails tended to change over the years, though the second mast often carried a square sail.

THE COMPASS

It is not known where or when it was discovered that the lodestone (a magnetized mineral composed of an iron oxide) aligns itself in a north-south direction, as does a piece of iron that has been magnetized by contact with a lodestone. Neither is it known where or when marine navigators first utilized these discoveries. Plausible records indicate that the Chinese were using the magnetic compass around 1100, western Europeans by 1187, Arabs by 1220, and Scandinavians by 1300. The device could have originated in each of these groups, or it could have been passed from one to the others. All of them had been making long voyages, relying on steady winds to guide them and sightings of the sun or a familiar star to inform them of any change. When the magnetic compass was introduced, it probably was used merely to check the direction of the wind when clouds obscured the sky.

The first mariner's compass may have consisted of a magnetized needle attached to a wooden splinter or a reed floating on water in a bowl. In a later version the needle was pivoted near its centre on a pin fixed to the bottom of

A magnetic compass. © Getty Images

the bowl. By the 13th century a card bearing a painted wind rose was mounted on the needle; the navigator could then simply read his heading from the card. So familiar has this combination become that it is called the compass, although that word originally signified the division of the horizon. The suspension of the compass bowl in gimbals (originally used to keep lamps upright on tossing ships) was first mentioned in 1537.

On early compass cards the north point was emphasized by a broad spearhead and the letter *T* for "tramontana," the name given to the north wind. About 1490 a combination of these evolved into the fleur-de-lis, still almost universally used. The east point, pointing toward the Holy Land, was marked with a cross; the ornament into which this cross developed continued on British compass cards well into the 19th century. The use of 32 points by sailors of northern Europe, usually attributed to Flemish compass makers, is mentioned by Geoffrey Chaucer in his *Treatise on the Astrolabe* (1391). It also has been said that the

Engineer's compass. © Corbis

navigators of Amalfi, Italy, first expanded the number of compass points to 32, and they may have been the first to attach the card to the needle.

During the 15th century it became apparent that the compass needle did not point true north from all locations but made an angle with the local meridian. This phenomenon was originally called by seamen the "northeasting" of the needle but is now called the "variation" or "declination." For a time, compass makers in northern countries mounted the needle askew on the card so that the fleur-de-lis indicated true north when the needle pointed to magnetic north. This practice died out about 1700 because it succeeded only for short voyages near the place where the compass was made. It caused confusion and difficulty on longer trips, especially in crossing the Atlantic to the American coast, where the declination was west instead of east as in Europe. The declination in a given location varies over time. For example, in northern Europe in the 16th century the magnetic north pole was east of true geographic north; in subsequent centuries it has drifted to the west.

Despite its acknowledged value, the magnetic compass long remained a fragile, troublesome, and unreliable instrument, subject to mysterious disturbances. The introduction of iron and then steel for hulls and engines in the 19th century caused further concern because it was well known that nearby ironwork would deflect the compass needle. In 1837 the British Admiralty set up a committee to seek rational methods of ensuring the accuracy of compasses installed on iron ships. In 1840 the committee introduced a new design that proved so successful that it was promptly adopted by all the principal navies of the world. Further refinements, aimed at reducing the effects of engine vibration and the shock of gunfire, continued throughout the century.

THE STEAMBOAT

The question of the invention of the steamboat raises fierce chauvinistic claims, particularly among the British, French, and Americans. However, there seems to be broad agreement that the first serious effort was carried out by a French nobleman, Claude-François-Dorothée, Marquis de Jouffroy d'Abbans, on the Doubs River at Baum-des-Dames in the Franche-Comté in 1776. This trial was not a success, but in 1783 Jouffroy carried out a second trial with a much larger engine built three years earlier at Lyon. This larger boat, the *Pyroscaphe*, was propelled by two paddle wheels, substituted for the two "duck's feet" used in the previous trial. The trial took place on the gentle River Saône at Lyon, where the overburdened boat of 327,000 pounds (147,000 kg) moved against the current for some 15 minutes before it disintegrated from the pounding of the engines. This was unquestionably the first steam-powered boat to operate. There were subsequent French experiments, but further development of the steamboat was impeded by the French Revolution.

In the eastern United States James Rumsey, the operator of an inn at the Bath Springs spa in Virginia (later West Virginia), sought to interest George Washington in a model steamboat he had designed. On the basis of Washington's support, Virginia and Maryland awarded Rumsey a monopoly of steam navigation in their territories.

At the same time, another American, John Fitch, a former clockmaker from Connecticut, began experimenting with his vision of a steamboat. After much difficulty in securing financial backers and in finding a steam engine in America, Fitch built a boat that was given a successful trial in 1787. By the summer of 1788 Fitch and his partner, Henry Voight, had made repeated trips on the Delaware

River as far as Burlington, 20 miles (30 km) above Philadelphia, the longest passage then accomplished by a steamboat.

British inventors were active in this same period. Both Rumsey and Fitch ultimately sought to advance their steamboats by going to England, and Robert Fulton spent more than a decade in France and Britain promoting first his submarine and later his steamboat. In 1788 William Symington, son of a millwright in the north of England, began experimenting with a steamboat that was operated at 5 miles (8 km) per hour, faster than any previous trials had accomplished. He later claimed speeds of 6½ and 7 miles (11 km) per hour, but his steam engine was thought too weak to serve, and for the time his efforts were not rewarded. In 1801 Symington was hired by Lord Dundas, a governor of the Forth and Clyde Canal, to build a steam tug; the *Charlotte Dundas* was tried out on that canal in 1802. It proved successful in pulling two 70-ton barges the 19 ½ miles (31.4 km) to the head of the canal in six hours. The governors, however, fearing bank erosion, forbade its use on that route, and British experiments failed to lead further for some years.

Fulton's Steamboat

Instead, Robert Fulton, an American already well known in Europe, began to gain headway in developing a steamboat. British historians have tended to deny his contributions and assign them to his supposed piracy of British inventions. It has been shown that he could not have pirated the plans of the *Charlotte Dundas*, but the record remains largely uncorrected. Fulton's "invention" of the steamboat depended fundamentally on his ability to make use of Scottish inventor James Watt's patents for the steam engine, as Fitch could not. Having experimented on steamboats for many years, by the first decade of the 19th century Fulton had determined that paddle wheels were the most efficient means of propelling a boat, a decision appropriate to the broad estuarine rivers of the Middle Atlantic states. Fulton had built and tested on Aug. 9, 1803, a steamboat that ran four times to the Quai de Chaillot on the Seine River in Paris. As it operated at no more than 2.9 miles (4.6 km) per hour—slower than a brisk walk—he considered these results at best marginal.

Fulton returned to the United States in December 1806 to develop a successful steamboat with his partner Robert Livingston. A monopoly on steamboating in New York state had been previously granted to Livingston, a wealthy Hudson Valley landowner and American minister to France. On Aug. 17, 1807, what was then called simply the "North River Steamboat" steamed northward on the Hudson from the state prison. After spending the night at Livingston's estate of Clermont (whose name has ever since erroneously been applied to the boat itself) the "North River Steamboat" reached Albany eight hours later after a run at an average speed of 5 miles (8 km) per hour (against the flow of the Hudson River). This was a journey of such length and relative mechanical success that there can be no reasonable question it was the first unqualifiedly successful steamboat trial. Commercial service began immediately, and the boat made one and a half round-trips between New York City and Albany each week. Many improvements were required in order to establish scheduled service, but from the time of this trial forward Fulton and Livingston provided uninterrupted service, added steamboats, spread routes to other rivers and sounds, and finally, in 1811, attempted to establish steamboat service on the Mississippi River.

The trial on the Mississippi was far from a success but not because of the steamboat itself. Fulton, Livingston, and their associate Nicholas Roosevelt had a copy of their Hudson River boats built in Pittsburgh as the *New Orleans*. In September 1811 it set sail down the Ohio River, making an easy voyage as far as Louisville, but as a deep-draft estuarine boat it had to wait there for the flow of water to rise somewhat. Finally, drawing no more than 5 inches (12 cm) less than the depth of the channel, the *New Orleans* headed downriver. In an improbable coincidence, the steamboat came to rest in a pool below the Falls of the Ohio just before the first shock was felt of the New Madrid earthquake, the most severe temblor ever recorded in the United States. The earthquake threw water out of the Ohio and then the Mississippi, filling the floodplain of those rivers, changing their channels significantly, and choking those channels with uprooted trees and debris. When the *New Orleans* finally reached its destination it was not sent northward again on the service for which it had been built. Steamboats used on the deeper and wider sounds and estuaries of the northeastern United States were found to be unsuited to inland streams, however wide. Eventually boats drawing no more than 9–12 inches (23–30 cm) of water proved to be successful in navigating the Missouri River westward into Montana and the Red River into the South; this pattern of steamboating spread throughout much of interior America, as well as the interior of Australia, Africa, and Asia.

THE RAILROAD

The earliest railroads reinforced transportation patterns that had developed centuries before. During the Middle Ages most heavy or bulky items were carried by water wherever possible. By the 16th century canal building was being widely used in Europe to integrate waterway systems based on natural streams, and during the Industrial Revolution canal networks became urgent necessities. In the 50 years after 1775 England and Wales were webbed with canals to provide reasonably inexpensive transport of coal. But in areas of concentrated industry in hilly country, such as around Birmingham and in the "Black Country" of England, or areas of heavy coal production in droughty uplands, as in western County Durham, the transporting of coal by water seemed impracticable.

The Plateway

A development of the late Middle Ages, the plateway suggested a means to make steam-powered land transport practicable. A plateway consisted of two parallel rails or plates on which wheels might run with somewhat reduced friction. The wheels were guided by a flange either on the rail or on the wheel. The plateways depended on gravity-induced movement or animal traction; it was logical to think of substituting steam-engine traction, but the shift had to await improvements in the steam engine. The weight-to-power ratio was unfavourable until 1804, when a Cornish engineer, Richard Trevithick, constructed a steam engine of his own design. In 1802 at Coalbrookdale in Shropshire he built a steam-pumping engine that operated at a pressure of 145 pounds per square inch (1,000 kilopascals). He mounted the high-pressure engine on a car with wheels set to operate on the rails of a cast-iron tramroad located at Pen-y-Darren, Wales.

The Stockton and Darlington Railway

George Stephenson was the son of a mechanic and, because of his skill at operating Newcomen engines, served as chief

The New Castle, built by Richard Trevithick in 1803, was the first locomotive to do actual work. Courtesy of CSX Transportation, Inc.

mechanic at the Killingworth colliery northwest of Newcastle upon Tyne, England. In 1813 he examined the first practical and successful steam locomotive, that of John Blenkinsop, and, convinced that he could offer improvements, designed and built the *Blücher* in 1814. Later he introduced the "steam blast," by which exhaust was directed up the chimney, pulling air after it and increasing the draft. His success in designing several more locomotives brought him to the attention of the planners of a proposed railway linking the port of Stockton with Darlington, 8 miles (12 km) inland, where investment in the Bishop Auckland coalfield of western County Durham was heavily concentrated and where there was agitation for improvement in the outward shipment of the increasing tonnages produced. A canal linking the cities had been proposed as early as 1769 but was rejected because of cost.

In 1818 the promoters settled on the construction of a railway, and in April 1821 parliamentary authorization was gained and George IV gave his assent.

While construction was under way on the 25-mile (40-km) single-track line it was decided to use locomotive engines as well as horse traction. Construction began on May 13, 1822, using both malleable iron rails (for two-thirds the distance) and cast iron and set at a track gauge of 4 feet, 8 inches (142 cm). This gauge was subsequently standardized, with one-half inch (11.2 cm) added at a date and for reasons unknown.

On Sept. 27, 1825, the Stockton and Darlington Railway was completed and opened for common carrier service between docks at Stockton and the Witton Park colliery in the western part of County Durham. It was authorized to carry both passengers and freight. From the beginning it was the first railroad to operate as a common carrier open to all shippers. Coal brought to Stockton for sale in the coastal trade dropped in price from 18 shillings to 12 shillings a ton. At that price the demand for coal was greater than the initial fabric of the Stockton and Darlington could handle.

This was an experimental line. Passenger service, offered by contractors who placed coach bodies on flatcars, did not become permanent until 1833, and horse traction was commonly used for passenger haulage at first. But after two years' operation the trade between Stockton and Darlington had grown tenfold.

THE LIVERPOOL AND MANCHESTER RAILWAY

The Liverpool and Manchester, Stephenson's second project, can logically be thought of as the first fully evolved railway to be built. It was intended to provide an extensive passenger service and to rely on locomotive traction alone.

THE *ROCKET*

The *Rocket* was a pioneer railway locomotive built by the English engineers George and Robert Stephenson. Following the success of the Stockton and Darlington Railway in 1825, the cities of Liverpool and Manchester decided to build a 40-mile (64-km) steam-operated line connecting them. George Stephenson was entrusted with constructing the line, but a competition was held to choose a locomotive. The Stephensons' *Rocket* won against three rivals, including an entry by John Ericsson, who later designed an armoured vessel called the *Monitor* for the federal forces during the American Civil War. For a short stretch the *Rocket* achieved a speed of 36 miles (58 km) per hour.

The Rainhill locomotive trials were conducted in 1829 to assure that those prime movers would be adequate to the demands placed on them and that adhesion was practicable. Stephenson's entry, the *Rocket*, which he built with his son, Robert, won the trials owing to the increased power provided by its multiple fire-tube boiler. The rail line began in a long tunnel from the docks in Liverpool, and the Edgehill Cutting through which it passed dropped the line to a lower elevation across the low plateau above the city. Embankments were raised above the level of the Lancashire Plain to improve the drainage of the line and to reduce grades on a gently rolling natural surface. A firm causeway was pushed across Chat Moss (swamp) to complete the line's quite considerable engineering works.

When the 30-mile (50-km) line was opened to traffic in 1830 the utility of railroads received their ultimate test. Though its cost had been more than £40,000 per mile and it could no longer be held that the railroad was a cheaper form of transportation than the canal, the Liverpool and Manchester demonstrated the railways' adaptability to diverse transportation needs and volumes. The opening of the line may fairly be regarded as the inauguration of the Railway Era, which continued until World War I.

THE BICYCLE

From Velocipedes to Boneshakers

Historians disagree about the invention of the bicycle, and many dates are challenged. It is most likely that no individual qualifies as the inventor and that the bicycle evolved through the efforts of many. The word *bicycle* came into use in Europe in 1868 to replace the cumbersome *vélocipède de pedale*. The first velocipede powered via pedals mounted on the front wheel was built in Paris during the early 1860s, but there is no conclusive evidence proving who conceived the idea of applying pedals to the front wheel or who actually did so. Pierre Michaux and his son Ernest presented their pedal-driven velocipede in the 1860s. The best evidence indicates that they built it in Paris in early 1864 (not 1861 or 1855, as stated in many histories), and a few more were built in 1865 and 1866. Some had malleable cast-iron frames, apparently in anticipation of large-scale production. Cranks and pedals were attached to the front wheel, which was 34 to 36 inches (86 to 91 cm) in diameter. The rear wheel was slightly smaller.

Michaux's role as the pioneer manufacturer of pedal bicycles is inextricably linked with the Olivier brothers, René and Aimé. In 1865 these two rich young men pedaled velocipedes more than 500 miles (800 km) from Paris to Marseille, and their subsequent enthusiasm for the new sport helped it to become a worldwide craze for the young, fit, and well-to-do. The brothers paid 50,000 francs for a 69 percent equity in Michaux, which then moved to a much larger factory. The first models had a serpentine-shaped malleable iron frame. Shortly thereafter the firm switched to a diagonal frame made of wrought iron, which quickly became the industry standard. By the fall of 1868, the new velocipede was a familiar sight across France, and sales reached new heights despite relatively high prices. In 1869 ball bearings and tension-spoked wheels were invented, and the freewheel (which allows coasting) was patented. The hard ride of wood-spoked wheels and iron rims gave early velocipedes the sobriquet of "boneshaker," but solid rubber tires and wire-spoked wheels helped soften the ride. In 1870, just as the boneshaker was developing into a practical bicycle called the "ordinary," the Franco-German War set back the French industry. Bicycle manufacture survived, but most subsequent developments took place in Britain.

THE PENNY-FARTHING

By the early 1870s bicycle technology and usage had come into its own. The crude boneshaker, based on wooden carriage technology, was replaced by the elegant "ordinary" bicycle. Hollow steel tubular frames and forks, quality ball bearings, tension-spoked wheels, steel rims, solid rubber tires, and standardized parts became common. James Starley's 1871 Ariel set the design standard for the ordinary bicycle. The Ariel had a 48-inch (122-cm) front wheel and a 30-inch (76-cm) rear wheel. Starley's prolific improvements for bicycles and tricycles over the next 10 years earned him the title "Father of the Cycle Trade." By 1874 the centre of the bicycle industry had shifted from Paris to Coventry, and England led technical development into the 20th century.

Two British companies exhibited bicycles at the 1876 Philadelphia Centennial Exposition. Albert E. Pope, a Boston industrialist, liked what he saw and began to import British ordinaries. By 1880 the Pope Manufacturing Co. was making the Columbia, a copy of the British Duplex Excelsior. This was the beginning of the American bicycle industry. The ordinary's cranks were directly connected to the front wheel, and its speed was limited by pedaling cadence and wheel diameter. Larger front wheels went faster and handled better on bad roads. Tension spoking allowed front wheels ranging from 40 to 60 inches (102 to 152 cm) in diameter, according to the owner's leg length. Though these high bicycles were called ordinaries, by the 1890s the term *penny-farthing* had come into use as a pejorative, comparing the front wheel to the large British penny and the rear wheel to the much smaller farthing (quarter-penny). Ordinaries typically weighed about 40 pounds (18 kg), but track-racing models could weigh as little as 16 pounds (7 kg). The ordinary was inherently unsafe. Mounting and dismounting required skill, and the rider sat almost directly over the large front wheel. From that position he could be pitched forward onto his head by road hazards. Also, the ordinary was slowed by reverse pressure on the pedals or by a lever-operated spoon brake, and severe braking or even hard back-pedaling could pitch the rider forward. Finally, the ordinary was expensive, so that most riders were athletic young men from the upper and middle classes.

The Safety Bicycle

As the ordinary was developing, numerous designs offered safer alternatives, including tricycles, gearing to allow smaller front wheels, and treadle drives to lower the pedals and the rider. These were called safety bicycles.

Chain-driven rear wheels were used on tricycles and prototype bicycles during the 1870s. Hans Renold invented the bush roller chain in Manchester, Eng., in 1880. This improved reliability and facilitated development of the safety bicycle.

The essential features of the safety bicycle were: spoked wheels roughly 30 inches (76 cm) in diameter, a chain-driven rear wheel with the front chainwheel roughly twice as large as the rear sprocket, a low centre of gravity, and direct front steering. Safety bicycles had decisive advantages in stability, braking, and ease of mounting. The first bicycle to provide all of these features and to achieve market acceptance was the 1885 Rover Safety designed by John Kemp Starley (James Starley's nephew). Prior to 1885 many alternative designs were called safety bicycles, but, after the Rover pattern took over the market in the late 1880s, safety bicycles were simply called bicycles. The last catalog year for ordinaries in England was 1892.

The early safety bicycles had solid rubber tires. In 1888 the pneumatic tire was introduced by John Boyd Dunlop, a Scottish veterinarian living in Belfast. These provided a more comfortable ride with greatly reduced rolling resistance. By 1893 virtually all new bicycles had pneumatic tires, which immensely increased their popularity. The pneumatic tire and the tension-spoked wheel did as much as the crank and pedal to establish the bicycle as a serious alternative to the horse.

THE AUTOMOBILE

Unlike many other major inventions, the original idea of the automobile cannot be attributed to a single individual. The idea certainly occurred long before it was first recorded in the *Iliad*, in which Homer (in Alexander Pope's translation) states that Vulcan in a single day made 20 tricycles, which:

Wondrous to tell instinct with spirit roll'd
From place to place, around the blest abodes,
Self-moved, obedient to the beck of gods.

Most authorities are inclined to honour Karl Benz and Gottlieb Daimler of Germany as the most important pioneer contributors to the gasoline-engine automobile. Benz ran his first car in 1885, Daimler in 1886. Although there is no reason to believe that Benz had ever seen a motor vehicle before he made his own, he and Daimler had been preceded by Étienne Lenoir in France and Siegfried Marcus in Austria, in 1862 and 1864–65, respectively, but neither Lenoir nor Marcus had persisted. Benz and Daimler did persist—indeed, to such purpose that their successor firm of Daimler AG can trace its origins back to the 1890s and claim, with the Peugeot SA firm of France, to be one of the oldest automobile-manufacturing firms in the world. Oddly, Benz and Daimler never met.

The Four-Stroke Gasoline Engine

The four-stroke principle upon which most modern automobile engines work was discovered by a French engineer, Alphonse Beau de Rochas, in 1862, a year before Lenoir ran his car from Paris to Joinville-le-Pont. The four-stroke cycle is often called the Otto cycle, after the German Nikolaus August Otto, who designed an engine on that principle in 1876. Beau de Rochas held prior patents, however, and litigation in the French courts upheld him. Lenoir's engine omitted the compression stroke of the Otto cycle; fuel was drawn into the cylinder on the intake stroke and fired by a spark halfway on the next reciprocal stroke.

The idea for Marcus's 1864–65 car apparently came to him by chance while he was considering the production of illumination by igniting a mixture of gasoline and air with a stream of sparks. The reaction was so violent that it occurred to him to use it as a power source. His first vehicle was a handcart marrying a two-cycle engine geared to the rear wheels without any intervening clutch. It was started by having a strong man lift the rear end while the wheels were spun, after which it ran for a distance of about 200 yards (about 180 m). Marcus's second model, the 1888–89 car, was sturdy and sufficiently well-preserved to make a demonstration run in the streets of Vienna in 1950, and again in 1987, at a rate of almost 3 miles (5 km) per hour. In 1898 the Austrian Automobile Club arranged an exhibition of motorcars, and Marcus was a guest of honour. Ironically, he denied interest in the idea of the automobile, calling it "a senseless waste of time and effort."

Karl Benz

Karl Benz was completely dedicated to the proposition that the internal-combustion engine would supersede the horse and revolutionize the world's transportation. He persisted in his efforts to build a gasoline-fueled vehicle in the face of many obstacles, including lack of money to the point of poverty and the bitter objections of his associates, who considered him unbalanced on the subject.

Benz ran his first car—a three-wheeler powered by a two-cycle, one-cylinder engine—on a happy and triumphant day early in 1885. He circled a cinder track beside his small factory, his workmen running beside the car and his wife running, too, clapping her hands. The little machine made four circuits of the track, stalling only twice before a broken chain stopped it. Even Max Rose, Benz's skeptical partner whose money had made the car possible, conceded that he was mildly impressed; but, like Siegfried Marcus, he remained convinced to the end of his association with Benz that there was no future in the horseless carriage.

Benz made his first sale to a Parisian named Émile Roger in 1888. Gradually, the soundness of his design and the quality and care that went into the material and the construction of his cars bore weight, and they sold well. That year he was employing some 50 workmen to build the tricycle car; in 1893 he began to make a four-wheeler.

In his way, Benz was almost as dogmatic and reactionary as Marcus had been; he objected to redesign of his original cars, and some authorities believe that he was never really convinced that his original concepts had been improved upon.

Gottlieb Daimler

Gunsmithing was Gottlieb Daimler's first vocation, and he showed marked talent, but he abandoned the trade to go to engineering school, studying in Germany, England, and France. In Germany he worked for various engineering and machining concerns, including the Karlsruhe Maschinenbaugesellschaft, a firm that much earlier had employed Benz.

In 1872 Daimler became technical director of Otto's firm, then building stationary gasoline engines. During the next decade, important work was done on the four-stroke engine. Daimler brought in several brilliant researchers, among them Wilhelm Maybach, but in 1882 both Daimler and Maybach resigned because of Daimler's conviction that Otto did not understand the potential of the internal-combustion engine. They set up a shop in Bad Cannstatt and built an air-cooled, one-cylinder engine. The first high-speed internal-combustion engine, it was designed to run at 900 revolutions per minute (rpm). For comparison, Benz's first tricycle engine had operated at only 250 rpm. Daimler and Maybach built a second engine and mounted it on a wooden bicycle fitted with an outrigger, which first ran on Nov. 10, 1885. The next year the first Daimler four-wheeled road vehicle was made: a carriage modified to be driven by a one-cylinder engine. Daimler appears to have believed that the first phase of the automobile era would be a mass conversion of carriages to engine drive; Benz apparently thought of the motorcar as a separate device. Daimler's licensees in France were René Panhard and Émile Levassor. In 1889 they entered the field independently, and the Panhard-Levassor designs of 1891–94 are of primary importance. They were true automobiles, not carriages modified for self-propulsion.

Daimler's 1889 car was a departure from previous practice. It was based on a framework of light tubing, it had the engine in the rear, its wheels were driven by a belt, and it was steered by a tiller. Remarkably, it had four speeds. This car had obvious commercial value, and in the following year the Daimler Motoren-Gesellschaft was founded. The British Daimler automobile was started as a manufactory licensed by the German company but later became quite independent of it. (To distinguish machines made by the two firms in the early years, the German cars are usually referred to as Cannstatt-Daimlers.) The Daimler and Benz firms were merged in 1926, and products thereafter have been sold under the name Mercedes-Benz.

THE AIRPLANE

Before recorded history humans knew of flight because they observed the birds, and in Greek mythology they sought to copy it, with grim consequences for Icarus. But experiments continued. In 1781 Karl Friedrich Meerwein, an architect to the prince of Baden, apparently succeeded in flying in an ornithopter (a flapping-wing machine, essentially a glider) at Giessen, Ger. This was one of the two main approaches to flying followed for a century and a quarter before directed human flight can be said to have been accomplished. The other approach was also observable in nature: in some conditions, such as that seen in the bubbles formed at the edge of waves breaking on a beach, enclosures of gas within a thin membrane would float off the Earth's surface, seeming to defy gravity. In time it was appreciated that different gases had different weights and that a lighter gas contained within a cell separated from a heavier general atmosphere formed the floating bubble buoyed upon the heavier gas. The gas-supported cell became a balloon, and as a source of flight it is a "lighter-than-air" craft, whereas the much refined successor of the ornithopter, which must do work to keep aloft, is a "heavier-than-air" craft.

Within a three-year period in the 1780s the two types had their first successful trials—fully documented in history for the balloon and more questionably so for the ornithopter. That flying machine, first "successfully" flown at Giessen, was a highly specialized form of glider, and only by using strong updrafts of air was it lifted off the surface. For most of the time until the Wright brothers' flight in 1903 the bubble was very much ahead in the competition for flight.

Early Experiments

The ornithopter in the 1780s had demonstrated that by applying a considerable amount of power to a machine of very light weight it should be possible to take off and fly above the Earth's surface in a heavier-than-air craft. This was accomplished by the "superlight" aircraft flights of the 1980s, including the successful crossing of the English Channel in a craft powered only by a single man's muscles.

Two problems arose: to find a favourable ratio between the weight of the vehicle and the power applied and to find a mechanical means to apply that power to lifting off the ground and achieving steerable forward motion. In 1799 the English physicist George Cayley worked out most of the aerodynamic theory. After Cayley's writing the ornithopter experiments were largely abandoned and replaced by trials of gliders, including Cayley's own in 1852–53. By the end of the 19th century the conditions were nearly ready for heavier-than-air flight. The development of the internal-combustion engine and of petroleum-based fuels (naphtha and gasoline) that were powerful in relation to weight meant that the problem of securing lift had essentially been solved. What remained were additional problems of applying that power to the vehicle. It is not without reason that the successful inventors of the airplane were two bicycle manufacturers from Dayton, Ohio: many of the problems of developing a rider-powered bicycle were reflected in shaping a self-powered heavier-than-air plane.

The Wright Brothers

Wilbur and Orville Wright in the course of their experiments came increasingly to consider Cayley's diagram of how a wing works, particularly the role played by the speed of the wind passing over the top of the wing. This led them to seek a site with a strong and persistent ambient wind (the Vogels Mountain where the 1781 ornithopter may have flown has just such a high ambient wind, as do the hills near Elmira, N.Y., and Fremont, Calif., classic gliding

THE WRIGHT FLYER OF 1903

The first powered airplane to demonstrate sustained flight under the full control of the pilot was assembled by Wilbur and Orville Wright in the autumn of 1903 at a camp at the base of the Kill Devil Hills, near Kitty Hawk, a village on the Outer Banks of North Carolina. After a first attempt failed on December 14, the machine was flown four times on December 17, to distances of 120, 175, 200, and 852 feet (36, 53, 61, and 260 m), respectively.

The 1903 Wright airplane was an extremely strong yet flexible braced biplane structure. Forward of the wings was a twin-surface horizontal elevator, and to the rear was a twin-surface vertical rudder. Wing spars and other long, straight sections of the craft were constructed of spruce, while the wing ribs and other bent or shaped pieces were built of ash. Aerodynamic surfaces were covered with a finely woven muslin cloth. The flyer was propelled by a four-cylinder gasoline engine of the Wrights' own design that developed some 12.5 horsepower after the first few seconds of operation. The engine was linked through a chain-drive transmission to twin contrarotating pusher propellers, which it turned at an average speed of 348 rotations per minute.

The pilot lay on the lower wing of the biplane with his hips positioned in a padded wooden cradle. A movement of the hips to the right or left operated the "wing-warping" system, which increased the angle of attack of the wings on one side of the craft and decreased it on the other, enabling the pilot to raise or lower the wing tips on either side in order to maintain balance or to roll into a turn. A small hand lever controlled the forward elevator, which provided pitch control and some extra lift. The rear rudder was directly linked to the wing-warping system in order to counteract problems of yaw produced by the warping of the wings.

> The 1903 machine was never flown after December 17. While sitting on the ground after the fourth flight, it was flipped by a gust of wind and badly damaged. Shipped back to Dayton, it was reassembled and repaired as needed for temporary exhibitions before being put on display at the Science Museum, London, in 1928. There it remained for 20 years, at the centre of a dispute between Orville Wright and the Smithsonian Institution in Washington, D.C., over claims that the Institution's third secretary, Samuel P. Langley, had constructed a machine capable of flight prior to the Wrights' flights of December 1903. The dispute ended with an apology from the Smithsonian in 1942, and the flyer was transferred permanently to the Institution's collection in 1948, several months after Orville's death.

courses). From the U.S. Weather Bureau the Wrights secured a list of windy sites in the United States, from which they chose the Outer Banks of North Carolina, specifically Kitty Hawk. On Kill Devil Hill there on Dec. 17, 1903, Orville Wright became the first man ever to fly in an aeroplane (as they were at first known), initially using as a frame a biplane of 40-foot 4-inch (1,230 km) wingspan and equipped with the 12-horsepower engine. He lifted off the ground in a 20–27-mile (32–43-km) wind and flew a distance of 120 feet (36 m) in 12 seconds. Having a strong wind certainly aided in that accomplishment, but the brothers soon demonstrated that such a wind was not absolutely essential.

After further experiments at Kitty Hawk they returned to Dayton to build a second plane, Flyer No. 2. Neither the balloons and dirigibles nor the earlier ornithopter and glider experiments had produced flight: what they had done was to harness the dynamics of the atmosphere to lift a craft off the ground, using what power (if any) they supplied primarily to steer. The Wrights initially used atmospheric dynamics to help in lifting the plane, but they subsequently demonstrated that they were able to lift a plane off the ground in still air.

In the long run their most significant invention was a way to steer the plane. After carefully watching a great number of birds, they became convinced that birds directed their flight by internally warping their wings, distorting them in one fashion or another. To do this in their plane, the Wrights constructed a ridged but distorted wing that might, through the use of wires fixed to the edge of the wing, be flexed to pass through the air in changing directions. This distortable wing was relatively misunderstood by other early plane experimenters.

During the summer of 1904 the Wrights made 105 takeoffs and managed to fly on a circular course up to 2.75 miles (4.5 km) for a sustained flight that lasted 5 minutes 4 seconds. Because they took a proprietary view of their invention, publicity about their work was minimal. After further trials in 1905 they stopped their experiments, using the time to obtain patents on their contribution. Only in 1908 did they break their secrecy when Wilbur Wright went to France to promote their latest plane.

THE SPACE LAUNCHER

The technology of rocket propulsion appears to have its origins in the period 1200–1300 CE in Asia, where the first "propellant" (a mixture of saltpetre, sulfur, and charcoal called black powder) had been in use for about 1,000 years for other purposes. As is so often the case with the development of technology, the early uses were primarily military. Powered by black powder charges, rockets served as bombardment weapons, culminating in effectiveness with the Congreve rockets (named for William Congreve, a British officer who was instrumental in their development) of the early 1800s. Performance of these early rockets was poor by modern standards because the only available propellant was black powder, which is not ideal for propulsion. Military use of rockets declined from 1815 to 1936 because of the superior performance of guns.

Tsiolkovsky, Goddard, and Oberth

During the period 1880–1930 the idea of using rockets for space travel grew in public interest. Stimulated by the conceptions of such fiction writers as Jules Verne, the Russian scientist Konstantin E. Tsiolkovsky worked on theoretical problems of propulsion-system design and rocket motion and on the concept of multistage rockets. Perhaps more widely recognized are the contributions of Robert H. Goddard, an American scientist and inventor who from 1908 to 1945 conducted a wide array of rocket experiments. He independently developed ideas similar to those of Tsiolkovsky about spaceflight and propulsion and implemented them, building liquid- and solid-propellant rockets. His developmental work included tests of the world's first liquid-propellant rocket in 1926. Goddard's many contributions to the theory and design of rockets earned him the title of "father of modern rocketry." A third pioneer, Hermann Oberth of Germany, developed much of the modern theory for rocket and spaceflight independent of Tsiolkovsky and Goddard. He not only provided inspiration for visionaries of spaceflight but also played a pivotal role in advancing the practical application of rocket propulsion that led to the development of rockets in Germany during the 1930s.

FROM THE V-2 TO THE ICBM

Due to the work of these early pioneers and a host of rocket experimenters, the potential of rocket propulsion was at least vaguely perceived prior to World War II, but there were many technical barriers to overcome. Development was accelerated during the late 1930s and particularly during the war years. The most notable achievements in rocket propulsion of this era were the German liquid-propellant V-2 rocket and the Me-163 rocket-powered airplane. (Similar developments were under way in other countries but did not see service during the war.) Myriad solid-propellant rocket weapons also were produced, and tens of millions were fired during combat operations by German, British, and U.S. forces. The main advances in propulsion that were involved in the wartime technology were the development of pumps, injectors, and cooling systems for liquid-propellant engines and high-energy solid propellants that could be formed into large pieces with reliable burning characteristics.

From 1945 to 1955 propulsion development was still largely determined by military applications. Liquid-propellant engines were refined for use in supersonic research aircraft, intercontinental ballistic missiles (ICBMs), and high-altitude research rockets. Similarly, developments in solid-propellant motors were in the areas of military tactical rocket applications and high-altitude research. Bombardment rockets, aircraft interceptors, antitank weapons, and air-launched rockets for air and surface targets were among the primary tactical applications. Technological advances in propulsion included the perfection of methods for casting solid-propellant charges, development of more energetic solid propellants, introduction of new structural and insulation materials in both

The Apollo 15 spacecraft lifts off from Cape Kennedy, Florida, U.S., atop a Saturn V three-stage rocket, July 26, 1971. A camera mounted at the mobile launch tower's 360-foot (110-m) level recorded this photograph. NASA

liquid and solid systems, manufacturing methods for larger motors and engines, and improvements in peripheral hardware (e.g., pumps, valves, engine-cooling systems, and direction controls). By 1955 most missions called for some form of guidance, and larger rockets generally employed two stages. While the potential for spaceflight was present and contemplated at the time, financial resources were directed primarily toward military applications.

The First "All-Civilian" System

The next decade witnessed the development of large solid-propellant rocket motors for use in ICBMs, a choice motivated by the perceived need to have such systems in ready-to-launch condition for long periods of time. This resulted in a major effort to improve manufacturing capabilities for large motors, lightweight cases, energetic propellants, insulation materials that could survive long operational times, and thrust-direction control. Enhancement of these capabilities led to a growing role for solid-rocket motors in spaceflight. Between 1955 and 1965 the vision of the early pioneers began to be realized with the achievement of Earth-orbiting satellites and manned spaceflight. The early missions were accomplished with liquid-propulsion systems adapted from military rockets. The first successful "all-civilian" system was the Saturn launch vehicle for the Apollo Moon-landing program, which used five 680,000-kilogram-thrust liquid-propellant engines in the first stage.

The Jetliner

From the very invention of flight at the beginning of the 20th century, military aircraft and engines generally led the way, and commercial aviation followed. At first this was also the case in the jet age, which began with the invention of jet engines under military sponsorship in the 1930s and '40s. By the late 20th century, however, commercial jet-engine technology had come to rival and sometimes even lead military technology in several areas of engine design. And, although it was not immediately evident, the invention of the jet engine had a far more significant social effect on the world through commercial aviation than through its military counterpart. Commercial jet aircraft have revolutionized world travel, opening up every corner of the world not just to the affluent but to the ordinary citizens of many countries.

Just as George Cayley and John Stringfellow of England, Lawrence Hargrave of Australia, Otto Lilienthal of Germany, and others had conducted experiments with flight in the years preceding Wilbur and Orville Wright's successful Wright flyer of 1903, so, too, were there many pioneers in the field of turbine engines before the almost simultaneous inventive successes of Frank Whittle of England and Hans von Ohain of Germany in the 1930s and '40s.

Whittle, von Ohain, and others met resistance to their ideas because conventional thinkers believed that the jet engine would produce too little power and consume too much fuel to be economically practical. It was not generally recognized that at higher altitudes the jet would produce more power with acceptable fuel efficiency. Understandably, even the most dedicated engine experts did not anticipate the rapid pace at which jet-engine performance would be improved.

It happened that the jet engine entered the propulsion scene at a time when conventional reciprocating engines and propellers were reaching their physical limits. Propellers were already encountering supersonic tip-speeds that destroyed their efficiency, and engines had grown so complex that additional horsepower in the 3,000–4,000 range depended on a large number of cylinders and complex supercharging that generated problems in operation and maintenance. With their continuous rotary motion, jet engines were mechanically simpler and smoother than reciprocating pistons with their rough pounding. Jet engines developed rapidly and by 1950 had reached levels of power that were impossible with piston engines.

It was not immediately obvious that the jet engine required major advances in airframe design and support facilities. First, airframes needed to be much larger to carry the additional passengers required to make jet aircraft economically sound. They would also have to be much stronger to accommodate the pressurized fuselage and the many transitions between low altitudes for takeoffs and landings and high altitudes for cruising. Another structural change was to sweep the wings back to reduce the drag increase associated with approaching supersonic flight. This was a possibility first elucidated by German engineer Adolph Buseman in 1935.

In Britain, the production of advanced commercial aircraft had been abandoned during the war, but a committee headed by aviation pioneer and former member of Parliament John Moore-Brabazon was established in 1943 to discuss postwar prospects of reviving the British air-transport industry. Among the suggestions was a specification for a transatlantic mailplane. De Havilland began design studies that led to the first flight of the D.H. 106 Comet jet airliner on July 27, 1949. The 36-seat Comet could fly at 500 miles (800 km) per hour for up to 1,500 miles (2,400 km). Boeing, Douglas, and Lockheed were stunned; though the Comet was considered too small and too short-ranged for American airline routes, they could offer no jet competitor. Britain's great lead faltered, however, when several Comets crashed, which led to its withdrawal from service in 1954. The later crashes were ultimately attributed to structural failure of the pressure cabin because of metal fatigue.

Boeing made a great advance with its revolutionary B-47 bomber, first flown on Dec. 17, 1947. The six-engine, swept-wing aircraft was purchased in large quantities (2,032) by the U.S. Air Force. This gave Boeing the engineering and financial basis to create the Model 367-80, a prototype for the 707 passenger plane. Although a tremendous gamble for Boeing, which for many years had been almost entirely a military supplier, the 707 was a commercial success after entering service in 1958. Douglas responded with its similar-looking DC-8. Both aircraft were larger (some configurations could carry more than 200 passengers) and faster (more than 600 miles [965 km] per hour) than the modified Comet 4 that began service on the New York to London route on Oct. 4, 1958.

France succeeded with its first effort at a jet airliner, creating the Sud-Est (later Aérospatiale) SE 210 Caravelle, a medium-range turbojet intended primarily for the continental European market. First flown on May 27, 1955, the Caravelle achieved sales of 282 aircraft, and a turbofan-powered variant was used for domestic routes by airlines in the United States—a marketing coup at the time. The Caravelle was the world's first airliner to have rear-mounted engines, a design feature that was adopted for some uses by all other major manufacturers.

The 1960s marked two stages in the jetliner's development. The first was the adoption of the turbofan engine. The turbofan gains economy by having much of its thrust pass around the engine core rather than through it. The second stage was the introduction of the wide-bodied, 400-seat Boeing 747 in 1969. This large, swift, long-ranged aircraft created a transportation revolution. Whereas air travel had once been confined to the affluent, it now became a mass-market conveyance.

In spite of the intense nature of the competition to build jet airliners, a new entrant appeared in the early 1970s following intense industrial and political negotiations. Airbus Industrie was co-owned by French, German, British, Spanish, Dutch, and Belgium companies and subcontracted many parts to still other countries. Established in December 1970 to build the Airbus A300 wide-bodied twin, the company was discounted at first as having little chance to compete. However, its aircraft were widely accepted, and a series of designs followed that established a family of aircraft that matched Boeing's offerings. With Airbus's introduction in 2005 of the A380, the world's largest airliner with a standard seating of 555, the company threatened to exceed Boeing.

GPS

A global positioning system (GPS) is any space-based radio-navigation system that broadcasts highly accurate navigation pulses to users on or near the Earth. In the United States' Navstar GPS, 24 main satellites in six orbits circle the Earth every 12 hours. Russia maintains a constellation called GLONASS (Global Navigation Satellite System), and in 2007 the European Union approved financing for the launch of 30 satellites to form its own version of GPS, known as Galileo, which is projected to be fully operational by 2013. China launched two satellites in 2000 and another in 2003 as part of a local navigation system first known as Beidou ("Big Dipper"), and in 2007 began launching a series of second-generation satellites, known as Beidou-2 or Compass. The constellation of 35 satellites is scheduled for completion in 2015.

A GPS receiver operated by a user on Earth measures the time it takes radio signals to travel from four or more satellites to its location, calculates the distance to each satellite, and from this calculation determines the user's longitude, latitude, and altitude. The U.S. Department of Defense originally developed the Navstar constellation for military use, but a less precise form of the service is available free of charge to civilian users around the globe. The basic civilian service will locate a receiver within 33 feet (10 m) of its true location, though various augmentation techniques can be used to pinpoint the location within less than 0.4 inch (1 cm). With such accuracy and the ubiquity of the service, GPS has evolved far beyond its original military purpose and has created a revolution in personal and commercial navigation. Battlefield missiles and artillery projectiles use GPS signals to determine their positions and velocities, but so do the U.S. space shuttle and the International Space Station as well as commercial jetliners and private airplanes. Ambulance fleets, family automobiles, and railroad locomotives benefit from GPS positioning, which also serves farm tractors, ocean liners, hikers, and even golfers. Many GPS receivers are no larger than a pocket calculator and are powered by disposable batteries, while GPS computer chips the size of a baby's fingernail have been installed in wristwatches, cellular telephones, and personal digital assistants.

The Navstar GPS system consists of three major segments: the space segment, the control segment, and the user segment. The space component is made up of the Navstar constellation in orbit around the Earth. The first satellite was an experimental Block I model launched in 1978. Nine more of these developmental satellites followed over the next decade, and 23 heavier and more-capable Block II production models were sent into space from 1989 to 1993. The launch of the 24th Block II satellite in

1994 completed the GPS constellation, which now consists of two dozen Block II satellites (plus three spares orbiting in reserve) marching in single file in six circular orbits around the Earth. The orbits are arranged so that at least five satellites are in view from most points on Earth at all times.

A typical Block II satellite weighs approximately 2,000 pounds (900 kg) and, with its solar panels extended, is about 56 feet (17 m) across. Its key elements are the winglike solar arrays that generate electrical power from sunlight, the 12 helical antennas that transmit navigation pulses to users on the ground, and its long, spearlike radio antenna that picks up instructions from control engineers. As a satellite coasts through its 12-hour orbit, its main body pivots continuously and the solar arrays swivel, keeping its navigation antennas pointing toward Earth's centre and its solar arrays aligned perpendicular to the Sun's rays.

The control segment consists of one Master Control Station at a U.S. Air Force base in Colorado and four additional unmanned monitoring stations positioned around the world—Hawaii and Kwajalein Atoll in the Pacific Ocean, Diego Garcia in the Indian Ocean, and Ascension Island in the Atlantic Ocean. Each monitoring station tracks all of the GPS satellites in its view to check for orbital changes. Every 18 months on average, the satellites within a given ring drift too far from their original configuration and must be nudged back with onboard thrusters fired by ground control.

The user segment consists of the millions of GPS receivers that pick up and decode the satellite signals. Hundreds of different types of GPS receivers are in use; some are designed for installation in automobiles, trucks, submarines, ships, aircraft, and orbiting satellites, whereas smaller models have been developed for personal navigation.

Chapter 3: Power and Energy

Fundamental to all inventions are power and energy, the harnessing of the capacity of the physical world to do work. The human race has applied its genius to using all sources of energy for a multitude of ends—among them, the combustion of fuel to heat homes and drive automobiles; the trapping of wind and water currents to turn wheels, raise levers, and generate electricity; and the directing of electricity to generate light or magnetic fields.

CONTROLLED FIRE

Fire is one of the human race's essential tools, control of which helped start it on the path toward civilization. The original source of fire undoubtedly was lightning, and such fortuitously ignited blazes long remained the only source of fire for humans and their ancestors. For some years Peking man (*Homo erectus*), about 500,000 BCE, was believed to be the earliest unquestionable user of fire. Evidence uncovered in Kenya in 1981 and in South Africa in 1988, however, suggests that the controlled use of fire by hominids may date from 1.4 to 1.5 million years ago. Not until about 7000 BCE did Neolithic man acquire reliable fire-making techniques, in the form of either drills, saws, and other friction-producing implements or of flint struck against pyrites. Even then it was more convenient to keep a fire alive permanently than to reignite it.

The first human beings to control fire gradually learned its many uses. Not only did they use fire to keep warm and cook their food; they also learned to use it in fire drives in hunting or warfare, to kill insects, to obtain berries, and to clear forests of underbrush so that game could be better seen and hunted. Eventually they learned that the burning of brush produced better grasslands and therefore more game.

With the achievement of agriculture in Neolithic times in the Middle East about 7000 BCE, there came a new urgency to clear brush and trees. The first agriculturists made use of fire to clear fields and to produce ash to serve as fertilizer. This practice, called slash-and-burn cultivation, persists in many tropical areas and some temperate zones today.

The step from the control of fire to its manufacture is great and required hundreds of thousands of years. The number and variety of inventions of such manufacture are difficult to imagine. Not until Neolithic times is there evidence that human beings actually knew how to produce fire. Whether a chance spark from striking flint against pyrites or a spark made by friction while drilling a hole in wood gave human beings the idea for producing fire is not known; but flint and pyrites, as well as fire drills, have been recovered from Neolithic sites in Europe.

Most widespread among prehistoric and later primitive peoples is the friction method of producing fire. The simple fire drill, a pointed stick of hard wood twirled between the palms and pressed into a hole on the edge of a stick of softer wood, is almost universal. The fire plow and the fire saw are variations on the friction method common in Oceania, Australia, and Indonesia. Mechanical fire drills were developed by the Eskimo, ancient Egyptians, Asian peoples, and a few Native American peoples. A fire piston that produced heat and fire by the compression of air in a small tube of bamboo was a complex device invented and used in southeastern Asia, Indonesia, and the Philippines. About 1800 a metal fire piston was independently invented in Europe. In 1827 the English chemist John Walker invented the friction match containing phosphorus sulfate, essentially the same as that which is in use today.

THE WATERWHEEL

The earliest machines were waterwheels, first used for grinding grain. They were subsequently adopted to drive sawmills and pumps, to provide the bellows action for furnaces and forges, to drive tilt hammers or trip-hammers for forging iron, and to provide direct mechanical power for textile mills. Until the development of steam power during the Industrial Revolution at the end of the 18th century, waterwheels were the primary means of mechanical power production, rivaled only occasionally by windmills. Thus, many industrial towns, especially in early America, sprang up at locations where water flow could be assured all year.

The oldest reference to a water mill dates to about 85 BCE, appearing in a poem by an early Greek writer celebrating the liberation from toil of the young women who operated the querns (primitive hand mills) for grinding corn. According to the Greek geographer Strabo, King Mithradates VI of Pontus in Asia used a hydraulic machine, presumably a water mill, by about 65 BCE.

Early vertical-shaft water mills drove querns where the wheel, containing radial vanes or paddles and rotating in a horizontal plane, could be lowered into the stream. The vertical shaft was connected through a hole in the stationary grindstone to the upper, or rotating, stone. The device spread rapidly from Greece to other parts of the world, because it was easy to build and maintain and could operate in any fast-flowing stream. It was known in China by the 1st century CE, was used throughout Europe by the end of the 3rd century, and had reached Japan by the year 610. Users learned early that performance could be improved with a millrace and a chute that would direct the water to one side of the wheel.

A horizontal-shaft water mill was first described by the Roman architect and engineer Vitruvius about 27 BCE. It consisted of an undershot waterwheel in which water enters below the centre of the wheel and is guided by a millrace and chute. The waterwheel was coupled with a right-angle gear drive to a vertical-shaft grinding wheel. This type of mill became popular throughout the Roman Empire, notably in Gaul, after the advent of Christianity led to the freeing of slaves and the resultant need for an alternative source of power. Early large waterwheels, which measured about 6 feet (1.8 m) in diameter, are estimated to have produced about three horsepower, the largest amount of power produced by any machine of the time. The Roman mills were adopted throughout much of medieval Europe, and waterwheels of increasing size, made almost entirely of wood, were built until the 18th century.

The first analysis of the performance of waterwheels was published in 1759 by John Smeaton, an English engineer. Smeaton built a test apparatus with a small wheel (its diameter was only 24 inches [61 cm]) to measure the effects of water velocity, as well as head and wheel speed. He found that the maximum efficiency (work produced divided by potential energy in the water) he could obtain was 22 percent for an undershot wheel and 63 percent for an overshot wheel (i.e., one in which water enters the wheel above its centre). In 1776 Smeaton became the first to use a cast-iron wheel, and two years later he introduced cast-iron gearing, thereby bringing to an end the all-wood construction that had prevailed since Roman times. Based on his model tests, Smeaton built an undershot wheel for the London Bridge waterworks that measured 15 feet (4.6 m) wide and that had a diameter of 32 feet (9.75 m). The results of Smeaton's experimental work came to be widely used throughout Europe for designing new wheels.

THE NORIA

The noria was an undershot waterwheel used to raise water in primitive irrigation systems. It was described by the Roman architect Vitruvius (c. 1st century BCE). As the noria turns, pots or hollow chambers on the rim fill when submerged and empty automatically into a trough when they reach or exceed the level of the centre of the wheel. In antiquity the wheels may have been as much as 40 feet (12 m) in diameter. Norias of the medieval period were even larger; the largest noria at Hamah, Syria, dates from 1000 CE and is 66 feet (20 m) in diameter. The wheel-turning force of the stream was sometimes augmented by humans or animals on a connected treadmill.

Early in the 19th century Jean-Victor Poncelet, a French mathematician and engineer, designed curved paddles for undershot wheels to allow the water to enter smoothly. His design was based on the idea that water would run up the surface of the curved vanes, come to rest at the inner diameter, and then fall away with practically no velocity. This design increased the efficiency of undershot wheels to 65 percent. At about the same time, William Fairbairn, a Scottish engineer, showed that breast wheels (i.e., those in which water enters at the 10- or two-o'clock position) were more efficient than overshot wheels and less vulnerable to flood damage. He used curved buckets and provided a close-fitting masonry wall to keep the water from flowing out sideways. In 1828 Fairbairn introduced ventilated buckets in which gaps at the bottom of each bucket allowed trapped air to escape. Other improvements included a governor to control the sluice gates and spur gearing for the power takeoff.

During the course of the 19th century, waterwheels were slowly supplanted by water turbines. Water turbines were more efficient; design improvements eventually made it possible to regulate the speed of the turbines and to run them fast enough to drive electric generators. This fact notwithstanding, waterwheels gave way slowly, and it was not until the early 20th century that they became largely obsolete.

THE WINDMILL

Windmills, like waterwheels, were among the original prime movers that replaced animal muscle as a source of power. They were used for centuries in various parts of the world, converting the energy of the wind into mechanical energy for grinding grain, pumping water, and draining lowland areas.

The first known wind device was described by Heron of Alexandria (*c.* 1st century CE). It was modeled on a water-driven paddle wheel and was used to drive a piston pump that forced air through a wind organ to produce sound. The earliest known references to wind-driven grain mills, found in Arabic writings of the 9th century CE, refer to a Persian millwright of 644 CE, although windmills may actually have been used earlier. These mills, erected near what is now the Iran–Afghanistan border, had a vertical shaft with paddlelike sails radiating outward and were located in a building with diametrically opposed openings for the inlet and outlet of the wind. Each mill drove a single set of millstones without gearing.

Windmills with vertical sails on horizontal shafts reached Europe through contact with the Arabs. Adopting the ideas from contemporary waterwheels, builders began to use fabric-covered, wood-framed sails located above the millstone, instead of a waterwheel below, to drive the grindstone through a set of gears. The whole mill with all its machinery was supported on a fixed post so that it could be rotated and faced into the wind. The millworks were initially covered by a boxlike wooden frame structure and later often by a "round-house," which also provided storage. A brake wheel on the shaft allowed the mill to be stopped by a rim brake. A heavy lever then had to be raised to release the brake, an early example of a fail-safe device. Mills of this sort first appeared in France in 1180, in areas of Syria under the control of the crusaders in 1190, and in England in 1191. The earliest known illustration is from the Windmill Psalter made in Canterbury, Eng., in the second half of the 13th century.

The large effort required to turn a post-mill into the wind probably was responsible for the development of the so-called tower mill in France by the early 14th century. Here, the millstone and the gearing were placed in a

Windmills in Spain. © Goodshoot/Jupiterimages

massive fixed tower, often circular in section and built of stone or brick. Only an upper cap, normally made of wood and bearing the sails on its shaft, had to be rotated. Such improved mills spread rapidly throughout Europe and later became popular with early American settlers.

The Low Countries of Europe, which had no suitable streams for waterpower, saw the greatest development of windmills. Dutch hollow post-mills, invented in the early 15th century, used a two-step gear drive for drainage pumps. An upright shaft that had gears on the top and bottom passed through the hollow post to drive a paddle-wheel-like scoop to raise water. The first wind-driven sawmill, built in 1592 in the Netherlands by Cornelis Cornelisz, was mounted on a raft to permit easy turning into the wind.

At first both post-mills and the caps of tower mills were turned manually into the wind. Later small posts were placed around the mill to allow winching of the mill with a chain. Eventually winches were placed into the caps of tower mills, engaged with geared racks and operated from inside or from the ground by a chain passing over a wheel. Tower mills had their sail-supporting or tail pole normally inclined at between 5° and 15° to the horizontal. This aided the distribution of the huge sail weight on the tail bearing and also provided greater clearance between the sails and the support structure. Windmills became progressively larger, with sails from about 55 to 78 feet (17 to 24 m) in diameter already common in the 16th century. The material of construction, including all gearing, was wood, although eventually brass or gunmetal came into use for the main bearings. Cast-iron drives were first introduced in 1754 by John Smeaton, the aforementioned English engineer. Little is known about the actual power produced by these mills. In all likelihood only from 10 to 15 horsepower was developed at the grinding wheels. A 50-horsepower mill was not built until the 19th century.

Post windmill with grinding machinery in mill housing, engraving from Agostino Ramelli's Li diverse et artificiose macchine, *1588. Courtesy of the trustees of the British Museum; photograph, J. R. Freeman & Co., Ltd.*

The maximum efficiency of large Dutch mills is estimated to have been about 20 percent.

Even though further improvements were made, especially in speed control, the importance of windmills as a major power producer began to decline after 1784, when the first flour mill in England successfully substituted a steam engine for wind power. Yet, the demise of windmills was slow; at one time in the 19th century there were as many as 900 corn (maize) and industrial windmills in the Zaan district of the Netherlands, the highest concentration known. Windmills persisted throughout the 19th century in newly settled or less-industrialized areas, such as the central and western United States, Canada, Australia, and New Zealand. They also were built by the hundreds in the West Indies to crush sugarcane.

The primary exception to the steady abandonment of windmills was resurgence in their use in rural areas for pumping water from wells. The first wind pump was introduced in the United States by David Hallay in 1854. After another American, Stewart Perry, began constructing wind pumps made of steel and equipped with metal vanes in 1883, this new and simple device spread around the world. Wind-driven pumps remain important today in many rural parts of the world. They continued to be used in large numbers, even in the United States, well into the 20th century until low-cost electric power became readily available in rural areas. Although rather inefficient, they are rugged and reliable, need little attention, and remain a prime source for pumping small amounts of water wherever electricity is not economically available.

THE STEAM ENGINE

The research of a number of scientists—especially those of Robert Boyle of England with atmospheric pressure, of

Otto von Guericke of Germany with a vacuum, and of the French Huguenot Denis Papin with pressure vessels — helped to equip practical technologists with the theoretical basis of steam power. Distressingly little is known about the manner in which this knowledge was assimilated by pioneers such as Thomas Savery and Thomas Newcomen, but it is inconceivable that they could have been ignorant of it. Savery took out a patent for a "new Invention for Raiseing of Water and occasioning Motion to all Sorts of Mill Work by the Impellent Force of Fire" in 1698 (No. 356). His apparatus depended on the condensation of steam in a vessel, creating a partial vacuum into which water was forced by atmospheric pressure.

Credit for the first commercially successful steam engine, however, must go to Newcomen, who erected his first machine near Dudley Castle in Staffordshire in 1712. It operated by atmospheric pressure on the top face of a piston in a cylinder, in the lower part of which steam was condensed to create a partial vacuum. The piston was connected to one end of a rocking beam, the other end of which carried the pumping rod in the mine shaft. Newcomen was a tradesman in Dartmouth, Devon, and his engines were robust but unsophisticated. Their heavy fuel consumption made them uneconomical when used where coal was expensive, but in the British coalfields they performed an essential service by keeping deep mines clear of water and were extensively adopted for this purpose. In this way the early steam engines fulfilled one of the most pressing needs of British industry in the 18th century. Although waterpower and wind power remained the basic sources of power for industry, a new prime mover had thus appeared in the shape of the steam engine, with tremendous potential for further development as and when new applications could be found for it.

Little development took place in the Newcomen atmospheric engine until James Watt patented a separate condenser in 1769, but from that point onward the steam engine underwent almost continuous improvements for more than a century. Watt's separate condenser was the outcome of his work on a model of a Newcomen engine that was being used in a University of Glasgow laboratory. Watt's inspiration was to separate the two actions of heating the cylinder with hot steam and cooling it to condense the steam for every stroke of the engine. By keeping the cylinder permanently hot and the condenser permanently cold, a great economy on energy used could be effected. This brilliantly simple idea could not be immediately incorporated in a full-scale engine because the engineering of such machines had up to that point been crude and defective. The backing of a Birmingham industrialist, Matthew Boulton, with his resources of capital and technical competence, was needed to convert the idea into a commercial success. Between 1775 and 1800, the period over which Watt's patents were extended, the Boulton and Watt partnership produced some 500 engines, which despite their high cost in relation to a Newcomen engine were eagerly acquired by the tin-mining industrialists of Cornwall and other power users who badly needed a more economic and reliable source of energy.

During the quarter of a century in which Boulton and Watt exercised their virtual monopoly over the manufacture of improved steam engines, they introduced many important refinements. Basically they converted the engine from a single-acting (i.e., applying power only on the downward stroke of the piston) atmospheric pumping machine into a versatile prime mover that was double-acting and could be applied to rotary motion, thus driving the wheels of industry. The rotary action engine was

THOMAS NEWCOMEN

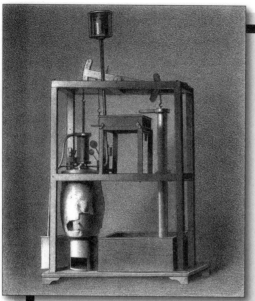

Model of a Newcomen steam engine, 1856.
© Photos.com/ Jupiterimages

English engineer and inventor Thomas Newcomen was born in 1663 in Dartmouth, Devon, Eng. As an ironmonger at Dartmouth, Newcomen became aware of the high cost of using the power of horses to pump water out of the Cornish tin mines. With his assistant John Calley (or Cawley), a plumber, he experimented for more than 10 years with a steam pump. It was superior to the crude pump of Thomas Savery. In Newcomen's engine the intensity of pressure was not limited by the pressure of the steam. Instead, atmospheric pressure pushed the piston down after the condensation of steam had created a vacuum in the cylinder.

As Savery had obtained a broad patent for his pump in 1698, Newcomen could not patent his atmospheric steam engine. Therefore, he entered into partnership with Savery. The first recorded Newcomen engine was erected near Dudley Castle, Staffordshire, in 1712.

Newcomen invented the internal-condensing jet for obtaining a vacuum in the cylinder and an automatic valve gear. By using steam at atmospheric pressure, he kept within the working limits of his materials. For a number of years his engine was used in the draining of mines and in raising water to power waterwheels. Newcomen died on Aug. 5, 1729, in London.

Newcomen steam engine, 1747. © Photos.com/Jupiterimages

quickly adopted by British textile manufacturer Sir Richard Arkwright for use in a cotton mill. And although the ill-fated Albion Mill, at the southern end of Blackfriars Bridge in London, was burned down in 1791—when it had been in use for only five years and was still incomplete—it demonstrated the feasibility of applying steam power to large-scale grain milling. Many other industries followed in exploring the possibilities of steam power, and it soon became widely used.

THE ELECTRIC BATTERY

The invention of the battery in 1800 made possible for the first time major advances in the theories of electric current and electrochemistry. Both science and technology developed rapidly as a direct result, leading some to call the 19th century the age of electricity.

GALVANI AND VOLTA

The development of the battery was the accidental result of biological experiments conducted by Luigi Galvani. Galvani, a professor of anatomy at the Bologna Academy of Science, was interested in electricity in fish and other animals. One day he noticed that electric sparks from an electrostatic machine caused muscular contractions in a dissected frog that lay nearby. At first, Galvani assumed that the phenomenon was the result of atmospheric electricity because similar effects could be observed during lightning storms. Later, he discovered that whenever a piece of metal connected the muscle and nerve of the frog, the muscle contracted. Although Galvani realized that some metals appeared to be more effective than others in producing this effect, he concluded incorrectly that the metal was transporting a fluid, which he identified with animal electricity, from the nerve to the muscle. Galvani's observations, published in 1791, aroused considerable controversy and speculation.

Alessandro Volta, a physicist at the nearby University of Pavia, had been studying how electricity stimulates the senses of touch, taste, and sight. When Volta put a metal coin on top of his tongue and another coin of a different metal under his tongue and connected their surfaces with a wire, the coins tasted salty. Like Galvani, Volta assumed that he was working with animal electricity until 1796 when he discovered that he could also produce a current when he substituted a piece of cardboard soaked in brine for his tongue. Volta correctly conjectured that the effect was caused by the contact between metal and a moist body. Around 1800 he constructed what is now known as a voltaic pile consisting of layers of silver, moist cardboard, and zinc, repeated in that order, beginning and ending with a different metal. When he joined the silver and the zinc with a wire, electricity flowed continuously through the wire. Volta confirmed that the effects of his pile were equivalent in every way to those of static electricity. Within 20 years, galvanism—as electricity produced by a chemical reaction was then called—became unequivocally linked to static electricity. More important, Volta's invention provided the first source of continuous electric current. This rudimentary form of battery produced a smaller voltage than the Leyden jar, a device invented by Pieter van Musschenbroek in 1745 that stored static electricity between two electrodes placed on the inside and outside of a jar. Volta's invention, however, was easier to use because it could supply a steady current and did not have to be recharged.

The controversy between Galvani, who mistakenly thought that electricity originated in the animal's nerve, and Volta, who realized that it came from the metal, divided scientists into two camps. Galvani was supported by Alexander von Humboldt in Germany, while Volta was backed by Charles-Augustin de Coulomb and other French physicists.

APPLICATIONS IN CHEMISTRY AND PHYSICS

Within six weeks of Volta's report, two English scientists, William Nicholson and Anthony Carlisle, used a chemical battery to discover electrolysis (the process in which an electric current produces a chemical reaction) and initiate the science of electrochemistry. In their experiment the two employed a voltaic pile to liberate hydrogen and oxygen from water. They attached each end of the pile to brass wires and placed the opposite ends of the wires into salt water. The salt made the water a conductor. Hydrogen gas accumulated at the end of one wire; the end of the other wire was oxidized. Nicholson and Carlisle discovered that the amount of hydrogen and oxygen set free by the current was proportional to the amount of current used. By 1809 the English chemist Sir Humphry Davy had used a stronger battery to free for the first time several very active metals—sodium, potassium, calcium, strontium, barium, and magnesium—from their liquid compounds. Michael Faraday, who was Davy's assistant at the time, studied electrolysis quantitatively and showed that the amount of energy needed to separate a gram of a substance from its compound is closely related to the atomic weight of the substance. Electrolysis became a method of measuring electric current, and the quantity of charge that releases a gram atomic weight of a simple element is now called a faraday in his honour.

Once scientists were able to produce currents with a battery, they could study the flow of electricity quantitatively. Because of the battery, the German physicist Georg Simon Ohm was able experimentally in 1827 to quantify precisely a problem that British scientist Henry Cavendish could only investigate qualitatively some 50 years earlier—namely, the ability of a material to conduct electricity. The result of this work—Ohm's law—explains how the resistance to the flow of charge depends on the type of conductor and on its length and diameter. According to Ohm's formulation, the current flow through a conductor is directly proportional to the potential difference, or voltage, and inversely proportional to the resistance—that is, $i = V/R$. Thus, doubling the length of an electric wire doubles its resistance, while doubling the cross-sectional area of the wire reduces the resistance by a half. Ohm's law is probably the most widely used equation in electric design.

THE ELECTRIC GENERATOR AND MOTOR

Electromagnetic technology began with English physicist and chemist Michael Faraday's discovery of induction in 1831. Faraday, the greatest experimentalist in electricity and magnetism of the 19th century and one of the greatest experimental physicists of all time, worked on and off for 10 years trying to prove that a magnet could induce electricity. In 1831 he finally succeeded by using two coils of wire wound around opposite sides of a ring of soft iron. The first coil was attached to a battery; when a current passed through the coil, the iron ring became magnetized. A wire from the second coil was extended to a compass needle 3 feet (1 m) away, far enough so that it was not affected directly by any current in the first circuit. When the first circuit was turned on, Faraday observed a momentary deflection of the compass needle and its immediate return to its original position. When the primary current was switched off, a similar deflection of the compass needle occurred but in the opposite direction. Building on this observation in other experiments, Faraday showed that changes in the magnetic field around the first coil are responsible for inducing the current in the second coil. He also demonstrated that an electric current can be induced by moving a magnet, by turning an electromagnet on and off, and even by moving an electric wire in the Earth's magnetic field. This demonstration showed that mechanical energy can be converted to electric energy. It provided the foundation for electric power generation, leading directly to the invention of the dynamo and the electric motor.

The early electric industry was dominated by the problem of generating electricity on a large scale. Within a year of Faraday's discovery, a small hand-turned generator in which a magnet revolved around coils was demonstrated in Paris. In 1833 there appeared an English model that featured the modern arrangement of rotating the coils in the field of a fixed magnet. By 1850 generators were manufactured commercially in several countries. Permanent magnets were used to produce the magnetic field in generators until the principle of the self-excited generator was discovered in 1866. (A self-excited generator has stronger magnetic fields because it uses electromagnets powered by the generator itself.) In 1870 Zénobe Théophile Gramme, a Belgian manufacturer, built the first practical generator capable of producing a continuous current. It was soon found that the magnetic field is more effective if the coil windings are embedded in slots in the rotating iron armature. The slotted armature, still in use today, was invented in 1880 by the Swedish engineer Jonas Wenström. Faraday's 1831 discovery of the principle of the alternating current (AC) transformer was not put to practical use until the late 1880s when the heated debate over the merits of direct-current and alternating-current systems for power transmission was settled in favour of the latter.

Electricity took on a new importance with the development of the electric motor. This machine, which converts electric energy to mechanical energy, has become an integral component of a wide assortment of devices ranging from kitchen appliances and office equipment to industrial robots and rapid-transit vehicles. Although the principle of the electric motor was devised by Faraday in 1821, no commercially significant unit was produced until 1873. In fact, the first important AC motor, built by the Serbian-American inventor Nikola Tesla, was not demonstrated in the United States until 1888. Tesla began producing his motors in association with the Westinghouse Electric Company a few years after direct current (DC) motors had been installed in trains in Germany and Ireland. By the end of the 19th century, the electric motor had taken a recognizably modern form. Subsequent improvements have rarely involved radically new ideas; however, the introduction of better designs and new bearing, armature, magnetic, and contact materials has resulted in the manufacture of smaller, cheaper, and more efficient and reliable motors.

THE INCANDESCENT LIGHTBULB

At first, the only serious consideration for electric power was arc lighting, in which a brilliant light is emitted by an electric spark between two electrodes. The carbon-arc electric light was demonstrated as early as 1808, and Faraday devised the first steam-powered electric generator to operate a large carbon-arc lamp for the Foreland Lighthouse in 1858. But the carbon-arc lamp was so bright and required so much power that it was never widely used, and it was limited to large installations like lighthouses, train stations, and department stores.

More practical lighting could be obtained from an incandescent lamp, any device that produced light by heating a suitable material to a high temperature. In 1801 the English chemist Sir Humphry Davy had demonstrated the incandescence of platinum strips heated in the open air by electricity, but the strips did not last long. Frederick de Moleyns of England was granted the first patent for an incandescent lamp in 1841; he used powdered charcoal heated between two platinum wires. Commercial development of an incandescent lamp, however, was delayed until a filament could be made that would heat to incandescence without melting and until a satisfactory vacuum tube could be built. The mercury pump, invented in 1865, provided an adequate vacuum, and a satisfactory carbon-filament bulb was developed independently by the English physicist Sir Joseph Wilson Swan in 1878 and the American inventor Thomas Edison the following year. By 1880 both had applied for patents for their incandescent lamps, and the ensuing litigation between the two men was resolved by the formation of a joint company in 1883. Edison has always received the major credit for inventing the lightbulb because of his development of the power lines and other equipment needed to establish the incandescent lamp in a practical lighting system.

The carbon-filament bulb was actually highly inefficient, but it banished the soot and fire hazards of coal-gas jets and soon gained wide acceptance. Indeed, thanks to the incandescent lamp, electric lighting became an accepted part of urban life by 1900. The carbon-filament bulb was eventually succeeded by the more efficient tungsten-filament incandescent bulb, developed by George Coolidge of the General Electric Company, which first appeared in 1908. In 1911 the drawn tungsten filament was introduced; in 1913 filaments were coiled, and bulbs were filled with inert gas. Nitrogen alone was used first, and, later, nitrogen and argon in proportions varied to suit the wattage. These steps increased efficiency. Beginning in 1925, bulbs were

THE FLUORESCENT LIGHT

A fluorescent light is an electric discharge lamp, cooler and more efficient than incandescent lamps, that produces light by the fluorescence of a phosphor coating. A fluorescent lamp consists of a glass tube filled with a mixture of argon and mercury vapour. Metal electrodes at each end are coated with an alkaline-earth oxide that gives off electrons easily. When current flows through the gas between the electrodes, the gas is ionized and emits ultraviolet radiation. The inside of the tube is coated with phosphors, substances that absorb ultraviolet radiation and fluoresce (reradiate the energy as visible light).

Because a fluorescent lightbulb does not provide light through the continual heating of a metallic filament, it consumes much less electricity than an incandescent bulb—only

Compact fluorescent lamps (bulbs). Encyclopaedia Britannica, Inc.

one-quarter the electricity or even less, by some estimates. However, up to four times the operating voltage of a fluorescent lamp is needed initially, when the lamp is switched on, in order to ionize the gas when starting. This extra voltage is supplied by a device called a ballast, which also maintains a lower operating voltage after the gas is ionized. In older fluorescent lamps the ballast is located in the lamp, separate from the bulb, and causes the audible humming or buzzing so often associated with fluorescent lamps. In newer compact fluorescent lights (CFLs), in which the fluorescent tube is coiled into a shape similar to an incandescent bulb, the ballast is nested into the cup at the base of the bulb assembly and is made of electronic components that reduce or eliminate the buzzing sound. The inclusion of a ballast in each individual bulb raises the cost of the bulb, but the overall cost to the consumer is still lower, owing to the reduced energy consumption and longer lifetime of the CFL. CFLs are rated by energy use (in watts) and light output (in lumens), frequently in specific comparison to incandescent bulbs. Specific CFLs are configured for use with dimmer switches, with three-way switches, and in recessed fixtures.

"frosted" on the inside with hydrofluoric acid to provide a diffused light instead of the glaring brightness of the unconcealed filament.

The double-coiled filament used today was introduced about 1930.

Since then, the filament lamp has become the principal form of electric lamp for domestic use. In 1938 General Electric and Westinghouse produced the first commercial fluorescent discharge lamps using mercury vapour and phosphor-coated tubes to enhance visible light output. Fluorescent tubes had roughly double the efficiency of tungsten lamps and were rapidly adopted for commercial and office use. In compact form they are finding growing use in homes as well. Other types of gaseous-discharge lamps—for example, using high-pressure mercury and sodium vapour—have been developed but have found only limited application in buildings; they are of such high intensity and marked colour that they are used mostly in high-ceilinged spaces and for exterior lighting.

THE STEAM TURBINE

A demand for power to generate electricity stimulated new thinking about the steam engine in the 1880s. The problem was that of achieving a sufficiently high rotational speed to make the dynamos that generated the electricity function efficiently. Such speeds were beyond the range of the normal reciprocating engine (i.e., with a piston moving backward and forward in a cylinder). Designers therefore began to investigate the possibilities of radical modifications to the reciprocating engine to achieve the speeds desired, or of devising a steam engine working on a completely different principle.

Early Precursors

Hints at that completely different principle might have been seen in the first device that can be classified as a reaction steam turbine—the aeolipile proposed by Heron of Alexandria, during the 1st century CE. In this device, steam was supplied through a hollow rotating shaft to a hollow rotating sphere. It then emerged through two opposing curved tubes, just as water issues from a rotating lawn sprinkler. The device was little more than a toy, since no useful work was produced.

Another steam-driven machine, described in 1629 in Italy, was designed in such a way that a jet of steam impinged on blades extending from a wheel and caused it to rotate by the impulse principle. Starting with a 1784 patent by James Watt, the aforementioned developer of the steam engine, a number of reaction and impulse turbines were proposed, all adaptations of similar devices that operated with water. None were successful except for the units built by William Avery of the United States after 1837. In one such Avery turbine two hollow arms, about 29 inches (75 cm) long, were attached at right angles to a hollow shaft through which steam was supplied. Nozzles at the outer end of the arms allowed the steam to escape in a tangential direction, thus producing the reaction to turn the wheel. About 50 of these turbines were built for sawmills, cotton gins, and woodworking shops, and at least one was tried on a locomotive. While the efficiencies matched those of contemporary steam engines, high noise levels, difficult speed regulation, and frequent need for repairs led to their abandonment.

Modern Steam Turbines

No further developments occurred until the end of the 19th century when various inventors laid the groundwork for the modern steam turbine, a design of such novelty that it constituted a major technological innovation. A steam turbine consists of a rotor resting on bearings and enclosed in a cylindrical casing. The rotor is turned by steam impinging against attached vanes or blades on which it exerts a force in the tangential direction. Thus a steam turbine could be viewed as a complex series of windmill-like arrangements, all assembled on the same shaft. In 1884 Sir Charles Algernon Parsons, a British engineer, recognized the advantage of employing a large number of stages in series, allowing extraction of the thermal energy in the steam in small steps. By passing steam through the blades of a series of rotors of gradually increasing size (to allow for the expansion of the steam) the energy of the steam was converted to very rapid circular motion, which was ideal for generating electricity. Parsons also developed the reaction-stage principle according to which a nearly equal pressure drop and energy release takes place in both the stationary and moving blade passages. In addition, he subsequently built the first practical large marine steam turbines.

During the 1880s Carl G. P. de Laval of Sweden constructed small reaction turbines that turned at about 40,000 revolutions per minute to drive cream separators. Their high speed, however, made them unsuitable for other commercial applications. De Laval then turned his attention to single-stage impulse turbines that used convergent-divergent nozzles. From 1889 to 1897 de Laval built many turbines with capacities from about 15 to several hundred horsepower. His 15-horsepower turbines were the first employed for marine propulsion (1892). C. E. A. Rateau of France first developed multistage impulse turbines during the 1890s. At about the same time, Charles G. Curtis of the United States developed the velocity-compounded impulse stage.

By 1900 the largest steam turbine-generator unit produced 1,200 kilowatts, and 10 years later the capacity of such machines had increased to more than 30,000 kilowatts. This far exceeded the output of even the largest steam engines, making steam turbines the principal prime movers in central power stations after the first decade of the 20th century. Following the successful installation of a series of 68,000-horsepower turbines in the transatlantic passenger liners *Lusitania* and *Mauretania*, launched in 1906, steam turbines also gained preeminence in large-scale marine applications, first with vessels burning fossil fuels and then with those using nuclear power. Steam generator pressures increased from about 1,000 kilopascals gauge in 1895 to 1,380 kilopascals gauge by 1919 and then to 9,300 kilopascals gauge by 1940. Steam temperatures climbed from about 356°F (180°C; saturated steam) to 599°F (315°C; superheated steam) and eventually to 950°F (510°C) over the same time period, while heat rates decreased from about 38,000 to below 10,000 Btus (40 to 10.55 megajoules) per kilowatt-hour.

THE GASOLINE ENGINE

The internal-combustion engine, which followed in the 19th century as an improvement over the steam engine for many applications, cannot be attributed to any single inventor. The piston, thought to date as far back as 150 BCE, was used by metalworkers in pumps for blowing air. The piston-and-cylinder system was basic to the steam engine, which brought the component to a high state of efficiency. The steam engine, however, suffered from low thermal efficiency, great weight and bulk, and inconvenience of operation, all of which were primarily traceable to the necessity of burning the fuel in a furnace separate from the engine. It became evident that a self-contained power unit was desirable.

As early as the 17th century, several experimenters first tried to use hot gaseous products to operate pumps. By 1820 an engine was built in England in which hydrogen-air mixtures were exploded in a chamber. The chamber was then cooled to create a vacuum acting on a piston. The sale of such gas engines began in 1823. They were heavy and crude but contained many essential elements of later, more successful devices. In 1824 the French engineer Sadi Carnot published his now classic pamphlet "Reflections on the Motive Power of Heat," which outlined fundamental internal-combustion theory. Over the next several decades inventors and engineers built engines that used pressure produced by the combustion of fuels rather than a vacuum and engines in which the fuel was compressed before burning. None of them succeeded in developing an operational system, however. Finally, in 1860 Étienne Lenoir of France marketed an engine that operated on illuminating gas and provided reasonably satisfactory service. The Lenoir engine was essentially a converted double-acting steam engine with slide valves for admitting gas and air and for discharging exhaust products. Although the Lenoir engine developed little power and utilized only about 4 percent of the energy in the fuel, hundreds of these devices were in use in France and Britain within five years. They were used for powering water pumps and printing presses and for completing certain other tasks that required only limited power output.

A major theoretical advance occurred with the publication in 1862 of a description of the ideal operating cycle of an internal-combustion engine. The author, the French engineer Alphonse Beau de Rochas, laid down the following conditions as necessary for optimum efficiency: maximum cylinder volume with minimum cooling surface, maximum rapidity of expansion, maximum ratio of expansion, and maximum pressure of the ignited charge. He described the required sequence of operations as (1) suction during an entire outstroke of the piston, (2) compression during the following instroke, (3) ignition of the charge at dead centre and expansion during the next outstroke (the power stroke), and (4) expulsion of the burned gases during the next instroke. The engine Beau de Rochas

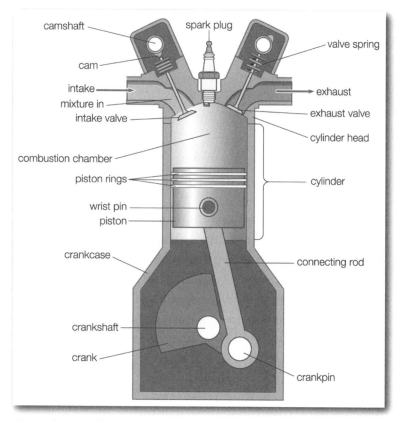

Typical piston-cylinder arrangement of a gasoline engine. Encyclopædia Britannica, Inc.

described thus had a four-stroke cycle, in contrast to the two-stroke cycle (intake-ignition and power-exhaust) of the Lenoir engine. Beau de Rochas never built his engine, and no four-stroke engine appeared for more than a decade. Finally, in 1876, the German engineer Nikolaus A. Otto built an internal-combustion unit based on Beau de Rochas's principle. (Otto's firm, Otto and Langen, had produced and marketed an improved two-stroke engine several years earlier.) The four-stroke Otto engine was an immediate success. In spite of its great weight and poor economy, nearly 50,000 engines with a combined capacity of about 200,000 horsepower were sold in 17 years, followed by the rapid development of a wide variety of engines of the same type. Manufacture of the Otto engine in the United States began in 1878, following the grant to Otto of a U.S. patent in 1877.

Eight years later Gottlieb Daimler and Wilhelm Maybach, former associates of Otto, developed the first successful high-speed four-stroke engine and invented a carburetor that made it possible to use gasoline for fuel. They employed their engine to power a bicycle (perhaps the world's first motorcycle) and later a four-wheeled carriage. At about the same time, another German mechanical engineer, Karl Benz, built a one-cylinder gasoline engine to power what is often considered the first practical automobile. The engines built by Daimler, Maybach, and Benz were fundamentally the same as today's basic gasoline engine.

THE JET ENGINE

A jet engine is a class of internal-combustion engines that propel aircraft by means of the rearward discharge of hot exhaust gases generated by burning fuel with air drawn in from the atmosphere. The prime mover of virtually all jet engines is a gas turbine. Variously called the core, gas producer, gasifier, or gas generator, the gas turbine converts the energy derived from the combustion of a liquid hydrocarbon fuel to mechanical energy in the form of a high-pressure, high-temperature airstream. The air ingested into the engine's inlet first is compressed by a turbocompressor, then flows into a combustion chamber, where a steady stream of fuel—in the form of liquid spray droplets and vapour or both—is introduced and burned. This gives rise to a continuous stream of high-pressure gas, which flows through a turbine to drive the compressor. The gas stream that exits from the gas generator after having been expanded through the turbine contains a considerable amount of surplus energy. This energy is harnessed by what is termed the propulsor (e.g., a jet nozzle, which accelerates the gas stream to a very high velocity as it leaves the engine) to generate a thrust with which to propel the aircraft.

Like many other inventions, jet engines were envisaged long before they became a reality. The earliest proposals were based on adaptations of piston engines and were usually heavy and complicated. The first successful experimental gas turbine using both rotary compressors and turbines was built in 1903 by Aegidus Elling of Norway. In this machine, part of the air leaving a centrifugal compressor was bled off for external power use. The remainder, which was required to drive the turbine, passed through a combustion chamber and then through a steam generator where the hot gas was partially cooled. This combustion gas was cooled further (by steam injected into it) to 750°F (400°C), the maximum temperature that Elling's radial-inflow turbine could handle. The earliest operational turbine of this type delivered 11 horsepower. Many subsequent improvements led to another experimental Elling turbine, which by 1932 could produce 75 horsepower. It employed a compressor with 71-percent efficiency and a turbine with an efficiency of 82 percent operating at an inlet temperature of 1,020°F (550°C). Norway's industry, however, was unable to capitalize on these developments, and no commercial units were built.

Also during the mid-1930s a group headed by Frank Whittle at the British Royal Aircraft Establishment (RAE) undertook efforts to design an efficient gas turbine for jet propulsion of aircraft. The unit produced by Whittle's group worked successfully during tests; it was determined

WHITTLE, VON OHAIN, AND THE FIRST JET AIRCRAFT

The German Heinkel He 178, the world's first successful jet airplane. U.S. Air Force

Beginning in the 1920s, steady advances in aircraft performance had been produced by improved structures and drag-reduction technologies and by more powerful, supercharged engines, but by the early 1930s it had become apparent to a handful of farsighted engineers that speeds would soon be possible that would exceed the capabilities of reciprocating engines and propellers. A few pioneers attacked the problem directly by conceiving a novel power plant, the turbojet.

While still a cadet at the Royal Air Force College, Cranwell, in 1928, Frank Whittle advanced the idea of replacing the piston engine and propeller with a gas turbine, and in the following year he conceived the turbojet, which linked a compressor, combustion chamber, and turbine in the same duct. In ignorance of Whittle's work, three German engineers independently arrived at the same concept: Hans von Ohain in 1933; Herbert Wagner, chief structural engineer for Junkers, a large German aircraft manufacturer, in 1934; and government aerodynamicist Helmut Schelp in 1937. Whittle had a running bench model by the spring of 1937, but backing from industrialist Ernst Heinkel gave von Ohain the lead.

The He 178, the first jet-powered aircraft, flew on Aug. 27, 1939, nearly two years before its British equivalent, the Gloster E.28/39, on May 15, 1941. Through an involved chain of events in which Schelp's intervention was pivotal,

Wagner's efforts led to the Junkers Jumo 004 engine. This became the most widely produced jet engine of World War II and the first operational axial-flow turbojet, one in which the air flows

The Heinkel He 178, the world's first turbojet-powered aircraft. Air Force Research Laboratory

straight through the engine. By contrast, the Whittle and Heinkel jets used centrifugal flow, in which the air is thrown radially outward during compression. Centrifugal flow offers advantages of lightness, compactness, and efficiency—but at the cost of greater frontal area, which increases drag, and lower compression ratios, which limit maximum power. Many early jet fighters were powered by centrifugal-flow turbojets, but, as speeds increased, axial flow became dominant.

that a pressure ratio of about 4 could be realized with a single centrifugal compressor running at roughly 17,000 revolutions per minute. Shortly after Whittle's achievement, another RAE group, led by A. A. Griffith and H. Constant, began developmental work on an axial-flow compressor. Axial-flow compressors, though much more complex and costly, were better suited for detailed blade-design analysis and could reach higher pressures and flow rates and, eventually, higher efficiencies than their centrifugal counterparts.

Independent parallel developments in Germany, initiated by Hans von Ohain working with the manufacturing firm of Ernst Heinkel, resulted in a fully operational jet aircraft engine that featured a single centrifugal compressor and a radial-inflow turbine. This engine was successfully tested in the world's first jet-powered airplane flight on Aug. 27, 1939. Subsequent German developments directed by Anselm Franz led to the Junkers Jumo 004 engine for the Messerschmitt Me 262 aircraft, which was first flown in 1942. In Germany as well as in Britain, the search for higher temperature materials and longer engine life was aided by experience gained in developing aircraft turbosuperchargers.

Before the end of World War II gas-turbine jet engines built by Britain, Germany, and the United States were flown in combat aircraft. Within the next few decades both propeller-driven gas-turbine engines (turboprops) and pure jet engines developed rapidly, with the latter assuming an ever larger role as airplane speeds increased.

Because of the significant advances in gas-turbine engine design in the years following World War II, it was expected that such systems would become an important prime mover in many areas of application. However, the high cost of efficient compressors and turbines, coupled with the continued need for moderate turbine-inlet temperatures, have limited the adoption of gas-turbine engines. Their preeminence remains assured only in the field of aircraft propulsion for medium and large planes that operate at either subsonic or supersonic speeds. As for electric power generation, large central power plants that use steam or hydraulic turbines are expected to continue to predominate. The prospects appear bright, nonetheless, for medium-sized plants employing gas-turbine engines in combination with steam turbines. Further use of gas-turbine engines for peak power production is likely as well. These turbine engines also remain attractive for small and medium-sized, high-speed marine vessels and for certain industrial applications.

THE NUCLEAR REACTOR

Soon after the discovery of nuclear fission was announced in 1939, it was also determined that the fissile isotope involved in the reaction was uranium-238 and that neutrons were emitted in the process. Newspaper articles reporting the discovery mentioned the possibility that a fission chain reaction could be exploited as a source of power. World War II, however, began in Europe in September of 1939, and physicists in fission research turned their thoughts to using the chain reaction in a bomb. In early 1940 the U.S. government made funds available for research that eventually evolved into the Manhattan Project. The Manhattan Project included work on uranium enrichment to procure uranium-235 in high concentrations and also research on reactor development. The goal was twofold: to learn more about the chain reaction for bomb design and to develop a way of producing a new element, plutonium, which was expected to be fissile and could be isolated from uranium chemically.

Reactor development was placed under the supervision of the leading experimental nuclear physicist of the era, Enrico Fermi. Fermi's project, begun at Columbia University and first demonstrated at the University of Chicago, centred on the design of a graphite-moderated reactor. It was soon recognized that heavy water was a better moderator and would be more easily used in a reactor, and this possibility was assigned to the Canadian research team since heavy-water production facilities already existed in Canada. Fermi's work led the way, and on Dec. 2, 1942, he reported having produced the first self-sustaining chain reaction. His reactor, later called Chicago Pile No. 1 (CP-1), was made of pure graphite in which uranium metal slugs were loaded toward the centre with uranium oxide lumps around the edges. This device had no cooling system, as it was expected to be operated for purely experimental purposes at very low power. CP-1 was subsequently dismantled and reconstructed at a new laboratory site in the suburbs of Chicago, the original headquarters of what is now Argonne National Laboratory. The device saw continued service as a research reactor until it was finally decommissioned in 1953.

On the heels of the successful CP-1 experiment, plans were quickly drafted for the construction of the first production reactors. These were the early Hanford reactors, which were graphite-moderated, natural uranium-fueled, water-cooled devices. As a backup project, a production reactor of air-cooled design was built at Oak Ridge, Tenn.; when the Hanford facilities proved successful, this reactor was completed to serve as the X-10 reactor at what is now Oak Ridge National Laboratory. Shortly after the end of World War II, the Canadian project succeeded in building a zero-power, natural uranium-fueled research reactor, the so-called ZEEP (Zero-Energy Experimental Pile). The first enriched-fuel research reactor was completed at Los Alamos, N.M., at about this time as enriched uranium-235 became available for research purposes. In 1947 a 100-kilowatt reactor with a graphite moderator and uranium metal fuel was constructed in England, and a similar one was built in France the following year.

In 1953 President Dwight D. Eisenhower of the United States announced the Atoms for Peace program. This program established the groundwork for a formal U.S. nuclear power program and expedited international cooperation on nuclear power.

The earliest U.S. nuclear power project had been started in 1946 at Oak Ridge, but the program was abandoned in 1948, with most of its personnel being transferred to the naval reactor program that produced the first nuclear-powered submarine, the *Nautilus*. After 1953 the U.S.

nuclear power program was devoted to the development of several reactor types, of which three ultimately proved to be successful in the sense that they remain as commercial reactor types or as systems scheduled for future commercial use. These three were the fast breeder reactor (now called LMR); the pressurized-water reactor; and the boiling-water reactor. The first LMR was the Experimental Breeder Reactor, EBR-I, which was designed at Argonne National Laboratory and constructed at what is now the Idaho National Engineering Laboratory near Idaho Falls, Idaho. EBR-I was an early experiment to demonstrate breeding, and in 1951 it produced electricity from nuclear heat for the first time. As part of the U.S. nuclear power program, a much larger experimental breeder, EBR-II, was developed and put into service (with power generation) in 1963. The principle of the boiling-water reactor was first demonstrated in a research reactor in Oak Ridge, but development of this reactor type was also assigned to Argonne, which built a series of experimental systems designated BORAX in Idaho. One of these, BORAX-III, became the first U.S. reactor to put power into a utility line on a continuous basis. A true prototype, the Experimental Boiling Water Reactor, was commissioned in 1957. The principle of the pressurized-water reactor had already been demonstrated in naval reactors, and the Bettis Atomic Power Laboratory of the naval reactor program was assigned to build a civilian prototype at Shippingport, Pa. This reactor, the largest of the power-reactor prototypes, is often hailed as the first commercial-scale reactor in the United States.

THE LASER

The laser is an outgrowth of a suggestion made by Albert Einstein in 1916 that under the proper circumstances atoms could release excess energy as light—either spontaneously or when stimulated by light. German physicist Rudolf Walther Ladenburg first observed stimulated emission in 1928, although at the time it seemed to have no practical use.

In 1951 Charles H. Townes, then at Columbia University in New York City, thought of a way to generate stimulated emission at microwave frequencies. At the end of 1953, he demonstrated a working device that focused "excited" ammonia molecules in a resonant microwave cavity, where they emitted a pure microwave frequency. Townes named the device a maser, for "microwave amplification by the stimulated emission of radiation." Aleksandr Mikhaylovich Prokhorov and Nikolay Gennadiyevich Basov of the P. N. Lebedev Physical Institute in Moscow independently described the theory of maser operation. For their work all three shared the 1964 Nobel Prize for Physics.

An intense burst of maser research followed in the mid-1950s, but masers found only a limited range of applications as low-noise microwave amplifiers and atomic clocks. In 1957 Townes proposed to his brother-in-law and former postdoctoral student at Columbia University, Arthur L. Schawlow (then at Bell Laboratories), that they try to extend maser action to the much shorter wavelengths of infrared or visible light. Townes also had discussions with a graduate student at Columbia University, Gordon Gould, who quickly developed his own laser ideas. Townes and Schawlow published their ideas for an "optical maser" in a seminal paper in the Dec. 15, 1958, issue of *Physical Review*. Meanwhile, Gould coined the word *laser* ("light amplification by the stimulated emission of radiation") and wrote a patent application. Whether Townes or Gould should be credited as the "inventor" of the laser thus became a matter of intense debate and led to years of litigation. Eventually, Gould received a series of four

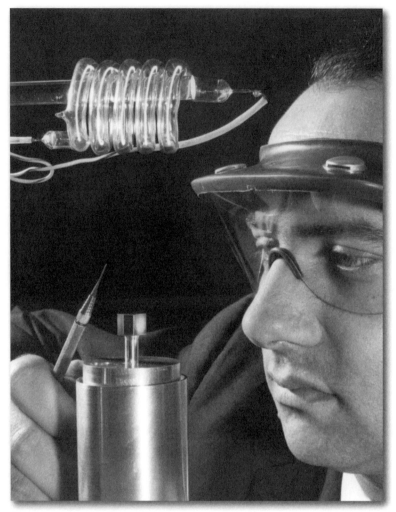

Theodore H. Maiman of Hughes Aircraft Company shows a cube of synthetic ruby crystal, the material at the heart of the first laser. Hughes Aircraft Company

patents starting in 1977 that earned him millions of dollars in royalties.

The Townes-Schawlow proposal led several groups to try building a laser. The Gould proposal became the basis of a classified military contract. Success came first to Theodore H. Maiman, who took a different approach at

Hughes Research Laboratories in Malibu, Calif. He fired bright pulses from a photographer's flash lamp to excite chromium atoms in a crystal of synthetic ruby, a material he chose because he had studied carefully how it absorbed and emitted light and calculated that it should work as a laser. On May 16, 1960, he produced red pulses from a ruby rod about the size of a fingertip. In December 1960 Ali Javan, William Bennett Jr., and Donald Herriott at Bell Labs built the first gas laser, which generated a continuous infrared beam from a mixture of helium and neon. In 1962 Robert N. Hall and coworkers at the General Electric Research and Development Center in Schenectady, N.Y., made the first semiconductor laser.

While lasers quickly caught the public imagination, perhaps for their similarity to the "heat rays" of science fiction, practical applications took years to develop. A young physicist named Irnee D'Haenens, while working with Maiman on the ruby laser, joked that the device was "a solution looking for a problem," and the line lingered in the laser community for many years. Townes and Schawlow had expected laser beams to be used in basic research and to send signals through air or space. Gould envisioned more powerful beams capable of cutting and drilling many materials. A key early success came in late 1963 when two researchers at the University of Michigan, Emmett Leith and Juris Upatnieks, used lasers to make the first three-dimensional holograms.

Helium-neon lasers were the first lasers with broad commercial applications. Because they could be adjusted to generate a visible red beam instead of an infrared beam, they found immediate use projecting straight lines for alignment, surveying, construction, and irrigation. Soon eye surgeons were using pulses from ruby lasers to weld detached retinas back in place without cutting into the eye. The first large-scale application for lasers was the laser scanner for automated checkout in supermarkets, which was developed in the mid-1970s and became common a few years later. Compact disc audio players and laser printers for personal computers soon followed.

Lasers have become standard tools in diverse applications. Laser pointers highlight presentation points in lecture halls, and laser target designators guide smart bombs to their targets. Lasers weld razor blades, write patterns on objects on production lines without touching them, remove unwanted hair, and bleach tattoos. Laser rangefinders in space probes profiled the surfaces of Mars and the asteroid Eros in unprecedented detail. In the laboratory, lasers have helped physicists to cool atoms to within a tiny fraction of a degree of absolute zero.

THE WIND TURBINE

Modern wind turbines extract energy from the wind, mostly for electricity generation, by rotation of a propeller-like set of blades that drive a generator through appropriate shafts and gears. The older term *windmill* is often still used to describe this type of device, although electric power generation rather than milling has become the primary application. As was noted earlier, windmills, together with waterwheels, were widely used to pump water, operate saws, and grind grain from the Middle Ages to the 19th century, at which time they were supplanted by steam engines and steam turbines. Windmills continued to be used for pumping water in rural areas through the 20th century, even as the internal-combustion engine and electricity provided more reliable and usually less expensive power. The best-known machines of this type were the so-called American farm windmills, which came into wide use during the 1890s. Such devices consist of a rotor, which may have up to 20 essentially flat sheet-metal blades and a tail vane that keeps the rotor facing into the wind by swiveling the entire rotor assembly. Governing is automatic and overspeeding is avoided by turning the wheel off the wind direction, thus reducing the effective sail area while keeping the speed constant. A typical pump can deliver about 10 gallons (38 litres) per minute to a height of 100 feet (30 m) at a wind velocity of 15 miles per hour (6.7 m per second).

The development of the electric generator in the 19th century aroused some interest in the wind as a "free" power source. The first windmill to drive a generator was built in 1890 by P. LaCour in Denmark, using patent sails and twin fantails on a steel tower. Adopting the ideas gained from airfoil and aircraft propeller designs, windmill designers and manufacturers began to replace broad windmill sails with a few slender propeller-like blades. In 1931 the first propeller wind turbine was erected in the Crimea. The Jacobs three-bladed windmill, used widely between 1930 and 1960, could deliver about one kilowatt of power at a wind speed of 14 miles per hour (6.25 m per second), a typical average wind velocity in the United States about 60 feet (18 m) above ground.

Today, wind turbines for electric power generation are most commonly propeller-type machines. Commercial wind turbines are made up of a blade or rotor and an enclosure called a nacelle that contains a drive train atop a tall tower. Large wind turbines (producing up to 1.8 megawatts of power) can have a blade length of about 130 feet (40 m) and be placed on towers about 260 feet (80 m) tall. Smaller turbines can be used to provide power to individual homes. Wind farms are areas where a number of wind turbines are grouped together, providing a larger total energy source.

By the early 21st century, wind was contributing slightly more than 1 percent of the world's total electricity, and electricity generation by wind has been increasing dramatically because of concerns over the cost of petroleum and the effects of fossil fuel combustion on the climate and environment. From 2004 to 2007, for example, total wind power increased from 59 to 95 gigawatts worldwide. Germany possesses the most installed wind capacity (16.6 gigawatts), and Denmark generates the largest percentage of its electricity from wind (nearly 20 percent).

Challenges to the large-scale implementation of wind energy include siting requirements such as wind availability, aesthetic and environmental concerns, and land availability. Wind farms are most cost-effective in areas with consistent strong winds; however, these areas are not necessarily near large population centres. Thus, power lines and other components of electrical distribution systems must have the capacity to transmit this electricity to consumers. In addition, since wind is an intermittent and inconsistent power source, storing power may be necessary. Public advocacy groups have raised concerns about the potential disruptions that wind farms

Windmills on a hillside in California are used to generate electricity. © MedioImages/Getty Images

may have on wildlife and overall aesthetics. For example, wind generators have been blamed for injuring and killing birds, though experts have shown that modern turbines have a small effect on bird populations. The National Audubon Society, a large environmental group based in the United States and focused on the conservation of birds and other wildlife, is strongly in favour of wind power, provided that wind farms are appropriately sited to minimize the impacts on migrating bird populations and important wildlife habitat.

THE SOLAR CELL

A solar cell is a device that directly converts the energy in light into electrical energy through the photovoltaic effect. The overwhelming majority of solar cells are fabricated from silicon—with increasing efficiency and lowering cost as the materials range from amorphous (noncrystalline) to polycrystalline to crystalline (single crystal) silicon forms. Unlike batteries or fuel cells, solar cells do not utilize chemical reactions or require fuel to produce electric power, and, unlike electric generators, they do not have any moving parts.

The development of solar cell technology stems from the work of the French physicist Antoine-César Becquerel in 1839. Becquerel discovered the photovoltaic effect while experimenting with a solid electrode in an electrolyte solution; he observed that voltage developed when light fell upon the electrode. About 50 years later, Charles Fritts constructed the first true solar cells using junctions formed by coating the semiconductor selenium with an ultrathin, nearly transparent layer of gold. Fritts's devices were very inefficient converters of energy; they transformed less than 1 percent of absorbed light energy into electrical energy. Though inefficient by today's standards, these early solar cells fostered among some a vision of abundant, clean power. In 1891 R. Appleyard wrote of:

> The blessed vision of the Sun, no longer pouring his energies unrequited into space, but by means of photo-electric cells . . ., these powers gathered into electrical storehouses to the total extinction of steam engines, and the utter repression of smoke.

By 1927 another metal-semiconductor-junction solar cell, in this case made of copper and the semiconductor copper oxide, had been demonstrated. By the 1930s both the selenium cell and the copper oxide cell were being employed in light-sensitive devices, such as photometers, for use in photography. These early solar cells, however, still had energy-conversion efficiencies of less than 1 percent. This impasse was finally overcome with the development of the silicon solar cell by Russell Ohl in 1941. Thirteen years later, aided by the rapid commercialization of silicon technology needed to fabricate the transistor, three other American researchers—Gerald Pearson, Daryl Chapin, and Calvin Fuller—demonstrated a silicon solar cell capable of a 6 percent energy-conversion efficiency when used in direct sunlight. By the late 1980s silicon cells, as well as cells made of gallium arsenide, with efficiencies of more than 20 percent had been fabricated. In 1989 a concentrator solar cell in which sunlight was concentrated onto the cell surface by means of lenses achieved an efficiency of 37 percent owing to the increased intensity of the collected energy. By connecting cells of different semiconductors optically and electrically in series, even higher efficiencies are possible, but at increased cost and added complexity. In general, solar cells of widely varying efficiencies and cost are now available.

Solar cells can be arranged into large groupings called arrays. These arrays, composed of many thousands of individual cells, can function as central electric power stations, converting sunlight into electrical energy for distribution to industrial, commercial, and residential users. Solar cells in much smaller configurations, commonly referred to as solar cell panels or simply solar panels, have been installed by homeowners on their rooftops to replace or augment their conventional electric supply. Solar cell panels also are used to provide electric power in many remote terrestrial locations where conventional electric power sources are either unavailable or prohibitively expensive to install. Because they have no moving parts that could need maintenance or fuels that would require replenishment, solar cells provide power for most space installations, from communications and weather satellites to space stations. (Solar power is insufficient for space probes sent to the outer planets of the solar system or into interstellar space, however, because of the diffusion of radiant energy with distance from the sun.) Another growing application of solar cells is in consumer products, such as electronic toys, handheld calculators, and portable radios. Solar cells used in devices of this kind may utilize artificial light (e.g., from incandescent and fluorescent lamps) as well as sunlight.

While total photovoltaic energy production is minuscule, it is likely to increase as fossil fuel resources shrink. In fact, the world's current energy consumption could be supplied by covering less than 1 percent of the Earth's surface with solar panels. The material requirements would be enormous but feasible, as silicon is the second most abundant element in the Earth's crust. These factors have led solar proponents to envision a future "solar economy" in which practically all of humanity's energy requirements are satisfied by cheap, clean, renewable sunlight.

THE FUEL CELL

A fuel cell is a device that converts the chemical energy of a fuel directly into electricity by electrochemical reactions. A fuel cell resembles a battery in many respects, but it can supply electrical energy over a much longer period of time. This is because a fuel cell is continuously supplied with fuel and air (or oxygen) from an external source, whereas a battery contains only a limited amount of fuel material and oxidant that is depleted with use. For this reason fuel cells have been used for decades in space probes, satellites, and manned spacecraft. Around the world thousands of stationary fuel cell systems have been installed in utility power plants, hospitals, schools, hotels, and office buildings for both primary and backup power. Many waste-treatment plants use fuel cell technology to generate power from the methane gas produced by decomposing garbage. Numerous municipalities lease fuel cell vehicles for public transportation and for use by service personnel. A small number of personal vehicles employing fuel cells have been sold on the market.

The general concept of a fuel battery, or fuel cell, dates back to the early days of electrochemistry. British physicist William Grove used hydrogen and oxygen as fuels catalyzed on platinum electrodes in 1839. During the late 1880s two British chemists—Carl Langer and German-born Ludwig Mond—developed a fuel cell with a longer service life by employing a porous nonconductor to hold the electrolyte. It was subsequently found that a carbon base permitted the use of much less platinum, and the German chemist Wilhelm Ostwald proposed, as a substitute for heat-engine generators, electrochemical cells in which carbon would be oxidized to carbon dioxide by oxygen. During the early years of the 20th century, Fritz Haber and Walther H. Nernst in Germany and Edmond Bauer in

France experimented with cells using a solid electrolyte. Limited success and high costs, however, suppressed interest in continuing developmental efforts.

From 1932 until well after World War II, British engineer Francis Thomas Bacon and his coworkers at the University of Cambridge worked on creating practical hydrogen-oxygen fuel cells with an alkaline electrolyte. Research resulted in the invention of gas-diffusion electrodes in which the fuel gas on one side is effectively kept in controlled contact with an aqueous electrolyte on the other side. By mid-century O. K. Davtyan of the Soviet Union had published the results of experimental work on solid electrolytes for high-temperature fuel cells and for both high- and low-temperature alkaline electrolyte hydrogen-oxygen cells.

The need for highly efficient and stable power supplies for space satellites and manned spacecraft created exciting new opportunities for fuel cell development during the 1950s and '60s. Molten carbonate cells with magnesium oxide pressed against the electrodes were demonstrated by J. A. A. Ketelaar and G. H. J. Broers of The Netherlands, while the very thin Teflon-bonded, carbon-metal hybrid electrode was devised by other researchers. Many other technological advances, including the development of new materials, played a crucial role in the emergence of today's practical fuel cells. Further improvements in electrode materials and construction, combined with the rising costs of fossil fuels, are expected to make fuel cells an increasingly attractive alternative power source, especially in Japan and other countries that have meager nonrenewable energy resources. At the beginning of the 21st century, many electrical-equipment manufacturers were developing power-generation equipment based on fuel cell technology.

Chapter 4: Building Construction and Civil Engineering

The building of great structures is a sign not only of human inventiveness but of organization. Over time we have gathered the intelligence, materials, and collective will to bridge chasms, erect soaring towers, domes, and vaults, and even create artificial controlled environments inside our own homes. Some of the greatest of those efforts are described here.

THE ARCH

In architecture and civil engineering, an arch is a curved member that is used to span an opening and to support loads from above. The arch can be said to have evolved from the post and lintel, and in turn it formed the basis for the evolution of the vault.

Arch construction depends essentially on the wedge. If a series of wedge-shaped blocks—i.e., ones in which the upper edge is wider than the lower edge—are set flank to flank, the result is an arch. These blocks are called voussoirs. Each voussoir must be precisely cut so that it presses firmly against the surface of neighbouring blocks and conducts loads uniformly. The central voussoir is called the keystone. The point from which the arch rises from its vertical supports is known as the spring, or springing line. During construction of an arch, the voussoirs require support from below until the keystone has been set in place; this support usually takes the form of temporary wooden centring. The curve in an arch may be semicircular, segmental (consisting of less than one-half of a circle), or pointed (two intersecting arcs of a circle); noncircular curves can also be used successfully.

Origins in the Post and Lintel

The basis for the evolution of all openings in construction is the post-and-lintel system, in which two upright members (posts, columns, piers) hold up a third member (lintel, beam, girder, rafter) laid horizontally across their top surfaces. The job of the lintel is to bear the loads that rest on it (and its own load) without deforming or breaking. Failure occurs only when the material is too weak or the lintel is too long. Lintels composed of materials that are weak in bending, such as stone, must be short, while lintels in materials that are strong in bending, such as steel, may span far greater openings. The job of the post is to support the lintel and its loads without crushing or buckling. Failure occurs, as in lintels, from excessive weakness or length, but the difference is that the material must be especially strong in compression. Stone, which has this property, is more versatile as a post than as a lintel; under heavy loads it is superior to wood but not to iron, steel, or reinforced concrete.

From prehistoric times to the Roman Empire, the post-and-lintel system was the root of architectural design. The interiors of Egyptian temples and the exteriors of Greek temples are delineated by columns covered by stone lintels. The Greeks opened their interior spaces by substituting wooden beams for stone, since the wood required fewer supports. The development of the arch and vault challenged the system but could not diminish its importance either in masonry construction or in wood framing, by its nature dependent on posts and beams.

Development of the Arch

The arch can be called a curved lintel. Early masonry builders could span only narrow openings because of the necessary shortness and weight of monolithic stone lintels. With the invention of the arch, two problems were solved: (1) wide openings could be spanned with small, light blocks, in brick as well as stone, which were easy to transport and to handle; and (2) the arch was bent upward to resist and to conduct into its supports the loads that tended to bend the lintel downward. Because the arch is curved, the upper edge has a greater circumference than the lower, so that each of its blocks must be cut in wedge shapes that press firmly against the whole surface of neighbouring blocks and conduct loads uniformly. This form creates problems of equilibrium that do not exist in lintels. The stresses in the arch tend to squeeze the blocks outward radially, and loads divert these outward forces downward to exert a resultant diagonal force, called thrust, which will cause the arch to collapse if it is not properly buttressed. So an arch cannot replace a lintel on two free-standing posts unless the posts are massive enough to buttress the thrust and to conduct it into the foundations (as in ancient Roman triumphal arches). Arches may rest on light supports, however, where they occur in a row, because the thrust of one arch counteracts the thrust of its neighbours, and the system will remain stable as long as the arches at either end of the row are buttressed by walls, piers, or earth. This system is used in such structures as arched stone bridges and ancient Roman aqueducts.

Arches were known in Egypt and Greece but were considered unsuitable for monumental architecture. In Roman times the arch was fully exploited in bridges, aqueducts, and large-scale architecture. In most cases the Romans did not use mortar, relying simply on the precision of their stone dressing. The Arabs popularized the pointed arch, and it was in their mosques that this form first acquired its religious connotations. Medieval Europe made great use of the pointed arch, which constituted a basic element in

THE VAULT

In essence, the vault is a structural member consisting of an arrangement of arches, usually forming a ceiling or roof. The basic barrel form, which appeared first in ancient Egypt and the Middle East, is in effect a continuous series of arches deep enough to cover a three-dimensional space. It exerts the same kind of thrust as the circular arch and must be buttressed along its entire length by heavy walls with limited openings. Roman architects discovered that two barrel vaults that intersected at right angles formed a groin vault, which, when repeated in series, could span rectangular areas of unlimited length. Because the groin vault's thrusts are concentrated at all four corners, its supporting walls need not be massive and require buttressing only where they support the vault.

Two large fragments of great concrete groin-vault buildings still survive from the late Roman Empire. The first of these is a portion of the Baths of Diocletian (c. 298–306) with a span of 85 feet (26 m); it was converted into the church of Santa Maria degli Angeli by Michelangelo in the 16th century. The other is the Basilica of Constantine (307–312 CE), also with a span of 85 feet. All of these buildings contained stone columns, but they were purely ornamental and could have been removed at will. The brick-faced concrete walls were left exposed on the exteriors, but the interiors were lavishly decorated with a veneer of thin slabs of coloured stone held in place by metal fasteners that engaged slots cut in the edges of the slabs, a technique still used in the 20th century.

Gothic architecture. In the late Middle Ages the segmental arch was introduced. This form and the elliptical arch had great value in bridge engineering because they permitted mutual support by a row of arches, carrying the lateral thrust to the abutments at either end of a bridge.

Steel, concrete, and laminated-wood arches of the 20th century changed the concept and the mechanics of arches. Their components are completely different from wedge-shaped blocks; they may be made entirely rigid so as to require only vertical support; they may be of hinged intersections that work independently, or they may be thin slabs or members (in reinforced concrete) in which stresses are so distributed that they add the advantages of lintels to those of arches, requiring only light supports. These innovations provide a great freedom of design and a means of covering great spans without a massive substructure.

BRICK

It was the cultures of the great river valleys — including the Nile, the Tigris and Euphrates, the Indus, and the Huang Ho — with their intensive agriculture based on irrigation, that developed the first communities large enough to be called cities. These cities were built with a new building technology, based on the clay available on the riverbanks. The packed clay walls of earlier times were replaced by those constructed of prefabricated units: mud bricks.

Bricks were made from mud and straw formed in a four-sided wooden frame, which was removed after evaporation had sufficiently hardened the contents. The bricks were then thoroughly dried in the sun. The straw acted as reinforcement, holding the brick together when the inevitable shrinkage cracks appeared during the drying process. The bricks were laid in walls with wet mud mortar or sometimes bitumen to join them together; openings were apparently supported by wooden lintels. In the warm, dry climates of the river valleys, weathering action was not a major problem, and the mud bricks were left exposed or covered with a layer of mud plaster. The roofs of these early urban buildings have disappeared, but it seems likely that they were supported by timber beams and were mostly flat, since there is little rainfall in these areas. Such mud brick or adobe construction is still widely used in the Middle East, Africa, Asia, and Latin America.

Later, about 3000 BCE in Mesopotamia, the first fired bricks appeared. Ceramic pottery had been developing in these cultures for some time, and the techniques of kiln-firing were applied to bricks, which were made of the same clay. Because of their cost in labour and fuel, fired bricks were used at first only in areas of greater wear, such as pavements or the tops of walls subject to weathering. They were used not only in buildings but also to build sewers to drain wastewater from cities. It is in the roofs of these underground drains that the first surviving true arches in brick are found, a humble beginning for what would become a major structural form. Corbel vaults and domes made of limestone rubble appeared at about the same time in Mesopotamian tombs. Corbel vaults are constructed of rows of masonry placed so that each row projects slightly beyond the one below, the two opposite walls thus meeting at the top. The arch and the vault may have been used in the roofs and floors of other buildings, but no examples have survived from this period. The well-developed masonry technology of Mesopotamia was used to build large structures of great masses of brick, such as the temple at Tepe Gawra and the ziggurats at Ur and Borsippa (Birs Nimrud), which were up to 87 feet (26 m) high. These symbolic buildings marked the beginnings of architecture in this culture.

DAMS

The oldest known dam in the world is a masonry and earthen embankment at Jawa in the Black Desert of modern Jordan. The Jawa Dam was built in the 4th millennium BCE to hold back the waters of a small stream and allow increased irrigation production on arable land downstream. Evidence exists of another masonry-faced earthen dam built about 2700 BCE at Sadd el-Kafara, about 19 miles (30 km) south of Cairo, Egypt. The Sadd el-Kafara failed shortly after completion when, in the absence of a spillway that could resist erosion, it was overtopped by a flood and washed away. The oldest dam still in use is a rockfill embankment about 20 feet (6 m) high on the Orontes River in Syria, built about 1300 BCE for local irrigation use.

The Assyrians, Babylonians, and Persians built dams between 700 and 250 BCE for water supply and irrigation. Contemporary with these was the earthen Ma'rib Dam in the southern Arabian Peninsula, which was more than 50 feet (15 m) high and nearly 1,970 feet (600 m) long. Flanked by spillways, this dam delivered water to a system of irrigation canals for more than 1,000 years. Remains of the Ma'rib Dam are still evident in present-day Ma'rib, Yemen. Other dams were built in this period in Sri Lanka, India, and China.

THE ROMANS

Despite their skill as civil engineers, the Romans' role in the evolution of dams is not particularly remarkable in terms of number of structures built or advances in height. Their skill lay in the comprehensive collection and storage of water and in its transport and distribution by aqueducts. At least two Roman dams in southwestern Spain, Proserpina and Cornalbo, are still in use, while the reservoirs of others have filled with silt. The Proserpina Dam, 40 feet (12 m) high, features a masonry-faced core wall of concrete backed by earth that is strengthened by buttresses supporting the downstream face. The Cornalbo Dam features masonry walls that form cells; these cells are filled with stones or clay and faced with mortar. The merit of curving a dam upstream was appreciated by at least some Roman engineers, and the forerunner of the modern curved gravity dam was built by Byzantine engineers in 550 CE at a site near the present Turkish-Syrian border.

Early Dams of East Asia

In East Asia, dam construction evolved quite independently from practices in the Mediterranean world. In 240 BCE a stone crib was built across the Jing River in the Gukou valley in China; this structure was about 100 feet (30 m) high and about 1,000 feet (300 m) long. Many earthen dams of moderate height (in some cases of great length) were built by the Sinhalese in Sri Lanka after the 5th century BCE to form reservoirs or tanks for extensive irrigation works. The Kalabalala Tank, which was formed by an earthen dam 79 feet (24 m) high and nearly 3.75 miles (6 km) in length, had a perimeter of 37 miles (60 km) and helped store monsoon rainfall for irrigating the country around the ancient capital of Anuradhapura. Many of these tanks in Sri Lanka are still in use today.

In Japan the Diamonike Dam reached a height of 105 feet (32 m) in 1128 CE. Numerous dams were also constructed in India and Pakistan. In India a design employing hewn stone to face the steeply sloping sides of earthen dams evolved, reaching a climax in the 10-mile-long

(16 km) Veeranam Dam in Tamil Nadu, built from 1011 to 1037 CE.

In Persia (modern-day Iran) the Kebar Dam and the Kurit Dam represented the world's first large-scale thin-arch dams. The Kebar and Kurit dams were built early in the 14th century by Il-Khanid Mongols; the Kebar Dam reached a height of 85 feet (26 m), and the Kurit Dam extended 210 feet (64 m) above its foundation. Remarkably, the Kurit Dam stood as the world's tallest dam until the beginning of the 20th century.

THE 15TH TO THE 18TH CENTURY IN EUROPE

In the 15th and 16th centuries, dam construction resumed in Italy and, on a larger scale, in Spain, where Roman and Moorish influence was still felt. In particular, the Tibi Dam across the Monnegre River in Spain, a curved gravity structure 138 feet (42 m) high, was not surpassed in height in western Europe until the building of the Gouffre d'Enfer Dam in France almost three centuries later. Also in Spain, the 75-foot-high (23 m) Elche Dam, which was built in the early 17th century for irrigation use, was an innovative thin-arch masonry structure. In the British Isles and northern Europe, where rainfall is ample and well distributed throughout the year, dam construction before the Industrial Revolution proceeded on only a modest scale in terms of height. Dams were generally limited to forming water reservoirs for towns, powering water mills, and supplying water for navigation canals. Probably the most remarkable of these structures was the 115-foot-high (35 m) earthen dam built in 1675 at Saint-Ferréol, near Toulouse, France. This dam provided water for the Midi Canal, and for more than 150 years it was the highest earthen dam in the world.

THE AQUEDUCT

An aqueduct (from the Latin *aqua*, "water," and *ducere*, "to lead") is a man-made conduit for carrying water. In a more restricted sense, aqueducts are structures used to conduct a water stream across a hollow or valley. In modern engineering *aqueduct* refers to a system of pipes, ditches, canals, tunnels, and supporting structures used to convey water from its source to its main distribution point.

Although the Romans are considered the greatest aqueduct builders of the ancient world, *qanāt* systems were in use in ancient Persia, India, Egypt, and other Middle Eastern countries hundreds of years earlier. These systems utilized slightly sloping tunnels driven into hillsides that contained groundwater. From the hillsides the water was conveyed by gravity in open channels to nearby towns or cities. The use of *qanāt*s became widespread throughout the region, and some are still in existence. Until 1933 the Iranian capital city, Tehrān, drew its entire water supply from a system of *qanāt*s.

Somewhat closer in appearance to the classic Roman structure was a limestone aqueduct built by the Assyrians around 691 BCE to bring fresh water to the city of Nineveh. Approximately 2,000,000 large blocks were used to make a water channel 30 feet (10 m) high and 900 feet (300 m) long across a valley.

The need to channel water supplies from distant sources was an outcome of the growth of urban communities. Among the most notable of ancient water conveyance systems are the aqueducts built between 312 BCE and 455 CE throughout the Roman Empire. Some of these impressive works are still in existence. The writings of Sextus Julius Frontinus (who was appointed superintendent of Roman aqueducts in 97 CE) provide information about the design and construction of the 11 major aqueducts that supplied Rome itself. Extending from a distant spring-fed area, a lake, or a river, a typical Roman aqueduct included a series of underground and aboveground channels. The longest was the Aqua Marcia, built in 144 BCE. Its source was about 23 miles (37 km) from Rome. The aqueduct itself was 57 miles (92 km) long, however, because it had to meander along land contours in order to maintain a steady flow of water. For about 50 miles (80 km) the aqueduct was underground in a covered trench, and only for the last 7 miles (11 km) was it carried aboveground on an arcade. In fact, most of the combined length of the aqueducts supplying Rome (about 260 miles [420 km]) was built as covered trenches or tunnels. When crossing a valley, aqueducts were supported by arcades comprising one or more levels of massive granite piers and impressive arches.

The aqueducts ended in Rome at distribution reservoirs, from which the water was conveyed to public baths or fountains. A few very wealthy or privileged citizens had water piped directly into their homes, but most of the

Roman aqueduct, Segovia, Spain. © Jupiterimages

people carried water in containers from a public fountain. Water was running constantly, the excess being used to clean the streets and flush the sewers. Ancient aqueducts and pipelines were not capable of withstanding much pressure. Channels were constructed of cut stone, brick, rubble, or rough concrete. Pipes were typically made of drilled stone or of hollowed wooden logs, although clay and lead pipes were also used.

Roman aqueducts were built throughout the empire, and their arches may still be seen in Greece, Italy, France, Spain, North Africa, and Turkey. As central authority fell apart in the 4th and 5th centuries, the systems also deteriorated. For most of the Middle Ages aqueducts were not used in western Europe, and people returned to getting their water from wells and local rivers. Modest systems sprang up around monasteries. By the 14th century, Bruges, with a large population for the time (40,000), had developed a system utilizing one large collecting cistern from which water was pumped, using a wheel with buckets on a chain, through underground conduits to public sites.

Modern aqueducts, although lacking the arched grandeur of those built by the Romans, greatly surpass the earlier ones in length and in the amount of water that they can carry. Aqueduct systems hundreds of miles long have been built to supply growing urban areas and crop-irrigation projects. The water supply of New York City comes from three main aqueduct systems that can deliver about 1.8 billion gallons (6.8 billion litres) of water a day from sources up to 120 miles (190 km) away. The aqueduct system in the state of California, however, is by far the largest in the world. It conveys water about 440 miles (700 km) from the northern (wetter) part of California into the southern (drier) part of the state, yielding more than 650 million gallons (2.5 billion l) of water a day.

THE ROMAN DOME

Under the Roman Republic, brickmaking, particularly in the region of Rome itself, became a major industry, and finally, under the Empire, it became a state monopoly. Brick construction was cheaper than stone due to the economies of scale in mass production and the lower level of skill needed to put it in place. The brick arch was adopted to span openings in walls, precluding the need for lintels. Mortar was at first the traditional mixture of sand, lime, and water, but, beginning in the 2nd century BCE, a new ingredient was introduced. The Romans called it *pulvis puteoli* after the town of Puteoli (modern Pozzuoli), near Naples, where it was first found. The material, formed in Mount Vesuvius and mined on its slopes, is now called pozzolana. When mixed with lime, pozzolana forms a natural cement that is much stronger and more weather-resistant than lime mortar alone and that will harden even underwater. Pozzolanic mortars were so strong and cheap, and could be placed by labourers of such low skill, that the Romans began to substitute them for bricks in the interiors of walls; the outer wythes of bricks were used mainly as forms to lay the pozzolana into place. Finally, the mortar of lime, sand, water, and pozzolana was mixed with stones and broken brick to form a true concrete, called *opus caementicium*. This concrete was still used with brick forms in walls, but soon it began to be placed into wooden forms, which were removed after the concrete had hardened.

The possibilities of plastic form suggested by this initially liquid material, which could easily assume curved shapes in plan and section, soon led to the creation of a series of remarkable interior spaces. These spaces, spanned by domes or vaults and uncluttered by the columns required by trabeated stone construction, showed the

THE PANTHEON

Interior of the Pantheon, Rome, *oil on canvas by Giovanni Paolo Panini, 1732. 119 × 98.4 cm.* In a private collection

The Pantheon, one of the greatest monumental buildings in Rome, was begun in 27 BCE by the statesman Marcus Vipsanius Agrippa, probably as a building of the ordinary Classical temple type—rectangular with a gabled roof supported by a colonnade on all sides. It was completely rebuilt by the emperor Hadrian sometime between 118 and 128 CE, and some alterations were made in the early 3rd century by the emperors Septimius Severus and Caracalla. It is a circular building of concrete faced with brick, with a great concrete dome rising from the walls and with a front porch of Corinthian columns supporting a gabled roof with triangular pediment. Beneath the porch are huge bronze double doors, 24 feet (7 m) high, the earliest known large examples of this type.

The Pantheon is remarkable for its size, its construction, and its design. Until modern times, the dome was the largest built, measuring about 142 feet (43 m) in diameter and rising to a height of 71 feet (22 m) above its base. There is no external evidence of brick arch support inside the dome, except in the lowest part, and the exact method of construction has never been determined. Two factors, however, are known to have contributed to its success: the excellent quality of the mortar used in the concrete and the careful selection and grading of the aggregate material, which ranges from heavy basalt in the foundations of the building and the lower part of the walls, through brick and tufa (a stone formed from volcanic dust), to the lightest of pumice toward the centre of the vault. In addition, the uppermost third of the drum of the walls, seen from the outside, coincides with the lower part of the dome, seen from the inside, and helps contain the thrust with internal brick arches. The drum itself is strengthened by huge brick arches and piers set above one another inside the walls, which are 20 feet (6 m) thick.

The porch is conventional in design, but the body of the building, an immense circular space lit solely by the light that floods through the 27-foot (8-m) "eye," or oculus, opening at the centre of the dome, was revolutionary. In contrast to the plain appearance of the outside, the interior of the building is lined with coloured marble, and the walls are marked by seven deep recesses, screened by pairs of columns whose modest size gives scale to the immensity of the rotunda. Rectangular coffers, or indentations, were cut in the ceiling, probably under Severus, and decorated with bronze rosettes and molding.

The Pantheon was dedicated in 609 CE as the Church of Santa Maria Rotonda, or Santa Maria ad Martyres, which it remains today. The bronze rosettes and moldings of the ceiling and other bronze embellishments have disappeared over time, and a frieze of stucco decoration was applied to the interior directly beneath the dome in the late Renaissance. Otherwise, the building exists entirely in its original form. The structure has been an enduring source of inspiration to architects since the Renaissance.

power of the imperial state. The first of these is the octagonal domed fountain hall of Nero's Golden House (64–68 CE), which is about 50 feet (15 m) in diameter with a large circular opening, or oculus, in the top of the dome. The domed form was rapidly developed in a series of imperial buildings that culminated in the emperor Hadrian's Pantheon of about 118–128 CE. This huge circular structure was entered from a portico of stone columns and was surmounted by a dome 142 feet (43 m) in diameter, lighted by an oculus at the top. The diameter of its circular dome remained unsurpassed until the 19th century.

In the late empire, concrete technology gradually disappeared, and even brickmaking ceased in western Europe. But significant developments in brick technology continued in the eastern Roman world, where the achievements of earlier periods in concrete were now duplicated in brickwork. The tomb of the emperor Galerius (now the Church of St. George) of about 300 CE at Salonika in Greece has a brick dome 80 feet (24 m) in diameter. It probably was the model for the climactic example of late Roman building, the great church of Hagia Sophia (532–537) in Constantinople, which features a central dome spanning 107 feet (32.6 m). Even Rome's great enemies, the Sāsānian Persians, built a large brick-vaulted hall in the palace at Ctesiphon (usually identified with Khosrow I [mid-6th century] but probably a 4th-century structure) with a span of 82 feet (25 m) by borrowing Roman methods. These late brick structures were the last triumphs of Roman building technology and would not be equaled for the next 900 years.

PLUMBING

One of the problems of every civilization in which the population has been centralized in cities and towns has been the development of adequate plumbing. The word *plumbing* comes from the Latin *plumbum*, which means "lead." One of the most important historical applications of lead was for the water pipes of Rome. The Romans provided generous water supplies for their cities; all of the supply systems worked by gravity and many of them used aqueducts and syphons. Although most people had to carry their water from public fountains, there was limited distribution of water to public buildings (particularly baths) and some private residences and apartment houses; private and semiprivate baths and latrines became fairly common. Lead pipes were fabricated in 10-foot (3-m) lengths and in as many as 15 standard diameters. Many of these pipes, still in excellent condition, have been uncovered in modern-day Rome and England.

The wastewater drainage system, on the other hand, was limited, with no treatment of sewage, which was simply discharged into an open, water-filled system of ditches that led to a nearby river. But even these fairly modest applications of public sanitation far exceeded those of previous cultures and would not be equaled until the 19th century.

With the coming of the Industrial Age, plumbing and sanitation systems in buildings advanced rapidly. Public water-distribution systems were the essential element; the first large-scale example of a mechanically pressurized water-supply system was the great array of waterwheels installed in the 17th century by French king Louis XIV at Marley on the Marne River to pump water for the fountains at Versailles, about 10 miles (18 km) away. The widespread use of cast-iron pipes in the late 18th century made higher pressures possible, and they were used by Napoleon in the first steam-powered municipal water supply for a section of Paris in 1812. Gravity-powered underground drainage systems were installed along with water-distribution networks in most large cities of the industrial world during the 19th century; sewage-treatment plants were introduced in the 1860s. Permanent plumbing fixtures appeared in buildings with water supply and drainage, replacing portable basins, buckets, and chamber pots.

Credit for the first flush toilet traditionally goes to Sir John Harington, an English courtier during the reign of Elizabeth I, who happened to be his godmother. Possibly around 1591 Harington installed a toilet for Elizabeth in her palace at Richmond, Surrey. In 1596, in *The Metamorphosis of Ajax* (a jakes; i.e., a toilet), he described his invention in terms more Rabelaisian than mechanical and was banished by Elizabeth, and he was never able to shed his reputation as Elizabeth's "saucy godson." Joseph Bramah invented the metal valve-type water closet as early as 1778, and other early lavatories, sinks, and bathtubs were of metal also; lead, copper, and zinc were all tried. The metal fixtures proved difficult to clean, however, and in England during the 1870s Thomas Twyford developed the first large one-piece ceramic lavatories as well as the ceramic washdown water closet. At first these ceramic fixtures were very expensive, but their prices declined until they became standard, and their forms remain largely unchanged today. The bathtub proved to be too large for brittle ceramic construction, and the porcelain-enamel cast-iron tub was devised about 1870. The double-shell built-in type still common today appeared about 1915.

THE PAVED ROAD

In Europe during the 17th and 18th centuries, technological improvements brought increased commercial travel, improved vehicles, and the breeding of better horses. These factors created an incessant demand for better roads, which up to this time had been built, with minor modifications, to the heavy Roman cross section. The typical Roman road was constructed on a raised embankment and composed of various layers of crushed rock and concrete, finally capped off by a wearing surface of large stone slabs — the total thickness varying from 3 to 6 feet (1 to 2 m). But in the last half of the 18th century a number of engineers appeared in France and Britain who would come to be known as the fathers of modern road building.

Pierre-Marie-Jérôme Trésaguet

In France, Pierre-Marie-Jérôme Trésaguet, an engineer from an engineering family, became in 1764 engineer of bridges and roads at Limoges and in 1775 inspector general of roads and bridges for France. In that year he developed an entirely new type of relatively light road surface, based on the theory that the underlying natural formation, rather than the pavement, should support the load. His standard cross section was 18 feet (5.5 m) wide and consisted of an 8-inch-thick (20 cm) course of uniform foundation stones laid edgewise on the natural formation and covered by a 2-inch (5-cm) layer of walnut-sized broken stone. This second layer was topped with a 1-inch (2.5-cm) layer of smaller gravel or broken stone. In order to maintain surface levels, Trésaguet's pavement was placed in an excavated trench — a technique that made drainage a difficult problem.

Thomas Telford

Thomas Telford, born of poor parents in Dumfriesshire, Scotland, in 1757, was apprenticed to a stonemason. Intelligent and ambitious, Telford progressed to designing bridges and building roads. He placed great emphasis on two features: (1) maintaining a level roadway with a maximum gradient of 1 in 30 and (2) building a stone surface capable of carrying the heaviest anticipated loads. His roadways were 18 feet (5.5 m) wide and built in three courses: (1) a lower layer, 7 inches (18 cm) thick, consisting of good-quality foundation stone carefully placed by hand (this was known as the Telford base), (2) a middle layer, also 7 inches thick, consisting of broken stone of 2-inch (5-cm) maximum size, and (3) a top layer of gravel or broken stone up to 1 inch (2.5 cm) thick.

John Loudon McAdam

The greatest advance came from John Loudon McAdam, born in 1756 at Ayr in Scotland. McAdam began his road-building career in 1787 but reached major heights after 1804, when he was appointed general surveyor for Bristol, then the most important port city in England. The roads leading to Bristol were in poor condition, and in 1816 McAdam took control of the Bristol Turnpike. There he showed that traffic could be supported by a relatively thin layer of small, single-sized, angular pieces of broken stone placed and compacted on a well-drained natural formation and covered by an impermeable surface of smaller stones. He had no use for the masonry constructions of his predecessors and contemporaries.

Drainage was essential to the success of McAdam's method, and he required the pavement to be elevated above the surrounding surface. The structural layer of broken stone was 8 inches (20 cm) thick and used stone of 2 to 3 inches (5 to 7.6 cm) maximum size laid in layers and compacted by traffic—a process adequate for the traffic of the time. The top layer was 2 inches thick, using three-fourths- to one-inch stone to fill surface voids between the large stones. Continuing maintenance was essential.

Although McAdam drew on the successes and failures of others, his total structural reliance on broken stone represented the largest paradigm shift in the history of road pavements. The principles of the "macadam" road are still used today.

REINFORCED CONCRETE

Beginning about the mid-19th century, the industrial world saw the reemergence of concrete in a new composite relationship with steel, creating a technology that would rapidly assume a major role in building construction. The first step in this process was the creation of higher-strength artificial cements. Lime mortar—made of lime, sand, and water—had been known since ancient times. It was improved in the late 18th century by the British engineer John Smeaton, who added powdered brick to the mix and made the first modern concrete by adding pebbles as coarse aggregate. Joseph Aspdin patented the first true artificial cement, which he called portland cement, in 1824; the name implied that it was of the same high quality as Portland stone. To make portland cement, Aspdin burned limestone and clay together in a kiln; the clay provided silicon compounds, which when combined with water formed stronger bonds than the calcium compounds of limestone. In the 1830s Charles Johnson, another British cement manufacturer, saw the importance of high-temperature burning of the clay and limestone to a white heat, at which point they begin to fuse. In this period, plain concrete was used for walls, and it sometimes replaced brick in floor arches that spanned between wrought-iron beams in iron-framed factories. Precast concrete blocks also were manufactured, although they did not effectively compete with brick until the 20th century.

The first use of iron-reinforced concrete was by the French builder François Coignet in Paris in the 1850s. Coignet's own all-concrete house in Paris (1862), the roofs and floors reinforced with small wrought-iron I beams, still stands. But reinforced concrete development actually began with the French gardener Joseph Monier's 1867 patent for large concrete flowerpots reinforced with a cage of iron wires. The French builder François Hennebique applied Monier's ideas to floors, using iron rods to reinforce concrete beams and slabs; Hennebique was the first to realize that the rods had to be bent upward to take negative moment near supports. In 1892 he closed his construction business and became a consulting engineer, building many structures with concrete frames composed of columns, beams, and slabs.

In the United States Ernest Ransome paralleled Hennebique's work, constructing factory buildings in concrete. High-rise structures in concrete followed the paradigm of the steel frame. Examples include the 16-story Ingalls Building (1903) in Cincinnati, which was 180 feet (54 m) tall, and the 11-story Royal Liver Building (1909), built in Liverpool by Hennebique's English representative, Louis Mouchel. The latter structure was Europe's first skyscraper, its clock tower reaching a height of 316 feet (95 m). Attainment of height in concrete buildings progressed slowly owing to the much lower strength and stiffness of concrete as compared with steel.

Concrete was also applied to long-span buildings, an early example being the Centennial Hall (1913) at Breslau, Ger. (now Wrocław, Pol.), by the architect Max Berg and the engineers Dyckerhoff & Widmann; its ribbed dome spanned 216 feet (65 m), exceeding the span of the Pantheon. More spectacular were the great airship hangars at Orly constructed by the French engineer Eugène Freyssinet in 1916; they were made with 3.5-inch-thick (9 cm) corrugated parabolic vaults spanning 266 feet (80 m) and pierced by windows. In the 1920s Freyssinet made a major contribution to concrete technology with the introduction of pretensioning. In this process, the reinforcing wires were stretched in tension, and the concrete was poured around them; when the concrete hardened, the wires were released, and the member acquired an upward deflection and was entirely in compression. When the service load was applied, the member deflected downward to a flat position, remaining entirely in compression, and it did not develop the tension cracks that plague ordinary reinforced concrete. Widespread application of pretensioning began after the end of World War II in 1945.

THE SUSPENSION BRIDGE

In a sense the suspension bridge is one of the oldest of engineering forms, bridges having been constructed by primitive peoples using vines for cables and mounting a path or roadway directly on the cables. A much stronger type of bridge was introduced in India about the 4th century CE that used cables of plaited bamboo and later of iron chain, with the roadway suspended from the cables — the true suspension bridge form. In modern times, the suspension bridge offered an economical solution to the problem of building long spans over navigable streams or at other sites where it is difficult to found piers in the stream. However, during the late 18th and early 19th centuries, British, French, American, and other engineers attempting to build suspension bridges encountered serious problems of stability and strength against wind forces and heavy loads; failures resulted from storms, heavy snows, and droves of cattle. Credit for solving the problem belongs principally to John Augustus Roebling, a German-born American engineer.

In 1841 Roebling established a factory for making rope out of iron wire, which he initially sold to replace the hempen rope used for hoisting cars over the portage

The George Washington Bridge, a vehicular suspension bridge across the Hudson River, U.S. © Jeffrey Sylvester/FPG

railway in central Pennsylvania. Later Roebling used wire ropes as suspension cables for bridges, and he developed the technique for spinning the cables in place rather than making a prefabricated cable that needed to be lifted into place. In 1855 Roebling completed a 821-foot-span (246 m) railway bridge over the Niagara River in western New York State. Wind loads were not yet understood in any theoretical sense, but Roebling recognized the practical need to prevent vertical oscillations. He therefore added numerous wire stays, which extended like a giant spiderweb in various directions from the deck to the valley below and to the towers above. The Niagara Bridge confounded nearly all the engineering judgment of the day, which held that suspension bridges could not sustain railway traffic. Although the trains were required to slow down to a speed of only 3 miles (5 km) per hour and repairs were frequent, the bridge was in service for 42 years, and it was replaced only because newer trains had become too heavy for it.

Roebling's Cincinnati Bridge (now called the John A. Roebling Bridge) over the Ohio River was a prototype for his masterful Brooklyn Bridge. When this 1,057-foot-span (317-m) span, iron-wire cable suspension bridge was completed in 1866, it was the longest spanning bridge in the world. Roebling's mature style showed itself in the structure's impressive stone towers and its thin suspended span, with stays radiating from the tower tops to control deck oscillations from wind loads.

Roebling died in 1869, shortly after work began on the Brooklyn Bridge, but the project was taken over and seen to completion by his son, Washington Roebling. Technically, the bridge overcame many obstacles through the use of huge pneumatic caissons, into which compressed air was pumped so that men could work in the dry; but, more important, it was the first suspension bridge on which steel wire was used for the cables. Every wire was galvanized to safeguard against rust, and the four cables, each nearly 16 inches (40 cm) in diameter, took 26 months to spin back and forth over the East River. After many political and technical difficulties and at least 27 fatal accidents, the 1,595-foot-span (479 m) bridge was completed in 1883 to such fanfare that within 24 hours an estimated quarter-million people crossed over it, using a central elevated walkway that John Roebling had designed for the purpose of giving pedestrians a dramatic view of the city.

DYNAMITE

Dynamite, a blasting explosive patented in 1867 by Alfred Nobel, was actually the second important invention by the Swedish chemist. Even today most experts regard Nobel's invention in 1865 of the blasting cap, a device for detonating explosives, as the greatest advance in the science of explosives since the discovery of black powder. After a number of attempts that were only partially successful, Nobel settled on a charge of mercury fulminate, which had been known for many years, in a copper capsule. With one or two minor changes, this blasting cap remained in general use until the 1920s.

The blasting cap provided a dependable means for detonating nitroglycerin and the many other high explosives that followed it. Nitroglycerin had been discovered by an Italian chemist, Ascanio Sobrero, in 1846. Because of the risks inherent in its manufacture and the lack of dependable means for its detonation, nitroglycerin was largely a laboratory curiosity until Nobel, together with his father, Immanuel Nobel, made extensive studies of its commercial potential in the years 1859–61. In 1862 they built a crude plant at Heleneborg, Sweden. Alfred, a chemist, was basically responsible for the design of this factory that was efficient and relatively safe considering the state of knowledge of the times. Nevertheless, it exploded in 1864 and killed, among others, Alfred's youngest brother Emil Oskar. Although deeply affected by the accident, Alfred continued work, at first on a barge that he moored in the middle of a lake. In 1865 he erected a plant at Krümmel, Ger., and another in Sweden at Vinterviken near Stockholm. A third plant was built a year later in Norway. Nobel was granted a patent for the manufacture and use of nitroglycerin in the United States, in 1866, and since importation on a large scale was impractical, he visited the United States in an effort to interest local capital. The victim of a number of unscrupulous businessmen, he finally sold his American holdings in 1885 for only $20,000.

Nobel coined the name *dynamite* from the Greek *dynamis*, "power." The basis for the invention was Nobel's discovery that kieselguhr, a porous siliceous earth, would absorb large quantities of nitroglycerin, giving an essentially dry and granular product that was much safer to handle and easier to use than nitroglycerin alone. Dynamite

No. 1, as Nobel called it, was 75 percent nitroglycerin and 25 percent guhr. Shortly after its invention, Nobel realized that guhr, an inert substance, not only contributed nothing to the power of the explosive but actually detracted from it because it absorbed heat that otherwise would have improved the blasting action. He turned, therefore, to active ingredients such as wood pulp for an absorbent and sodium nitrate for an oxidizing agent. By varying the ratio of nitroglycerin to these "dopes," as they came to be called, Nobel not only improved the efficiency of dynamite but also was able to prepare it in varying strengths, termed straight dynamites. Thus 40 percent straight dynamite contained 40 percent nitroglycerin and 60 percent dope.

Nobel's next outstanding contribution was his invention of gelatinous dynamites in 1875. There is a legend that he hurt a finger and used collodion, a solution of relatively low nitrogen content nitrocellulose in a mixture of ether and alcohol, to cover the wound. Later, unable to sleep because of the pain, Nobel went to the laboratory to find out what effect collodion would have on nitroglycerin. To his great satisfaction, he found that after evaporation of the solvents, there remained a tough, plastic material. He discovered that he could duplicate this by the direct addition of 7 to 8 percent of collodion-type nitrocotton to nitroglycerin and that lesser quantities of nitrocotton decreased the viscosity and enabled him to add other active ingredients. He called the original material *blasting gelatin* and the dope mixtures *gelatin dynamites*. The principal advantages of these products were their high water resistance and greater blasting action power than the comparable dynamites. This added power resulted from a combination of higher density and a degree of plasticity that allowed complete filling of the borehole (the hole that was bored in the coal seam or elsewhere for implantation of the explosive).

Three tunnels stand out as benchmarks in the history of the use of explosives and mark the transition from the older black powder to Nobel's dynamite: first is Mont Cenis, an 8-mile (13-km) railway tunnel driven through the Alps between France and Italy in 1857–71, the largest construction job with black powder up to that time; second was the 4-mile (6.4-km) Hoosac, also a railway project, during the construction of which (1855–66) nitroglycerin first replaced black powder in large-scale construction; third was the Sutro mine development tunnel in Nevada (1864–74), where the switch from nitroglycerin to dynamite for this type of work started.

THE SKYSCRAPER

Beginning about 1880, the mass production of steel and the growing supply of electricity transformed building technology. Steel was first made in large quantities for railroad rails. Rolling of steel rails (which was adapted from wrought-iron rolling technology) and other shapes such as angles and channels began about 1870; it made a much tougher, less brittle metal. Steel was chosen as the principal building material for two structures built for the Paris Exposition of 1889: the Eiffel Tower and the Gallery of Machines. Gustave Eiffel's tower was 1,000 feet (300 m) high, and its familiar parabolic curved form has become a symbol of Paris itself; its height was not exceeded until the topping off of the 1,046-foot-tall (318.8 m) Chrysler Building in New York City in 1929. The Gallery of Machines was designed by the architect C.-L.-F. Dutert and the engineer Victor Contamin with great three-hinged arches spanning 380 feet (114 m) and extending more than 1,400 feet (420 m). Its glass-enclosed clear span area of 536,000 square feet (48,727 sq m) has never been equaled; in fact, it was so large that no regular use for it could be

The Seagram Building, New York City, by Ludwig Mies van der Rohe and Philip Johnson, 1956–58. Photo Media, Ltd.

THE GLASS CURTAIN WALL

The second great age of high-rise buildings began after the end of World War II. It was an optimistic time with declining energy costs, and architects embraced the concept of the tall building as a glass prism. This idea had been put forward by the architects Le Corbusier and Ludwig Mies van der Rohe in their visionary projects of the 1920s. These designs employed the glass curtain wall, a non-load-bearing "skin" attached to the exterior structural components of the building. The earliest all-glass curtain wall, which was only on a single street facade, was that of the Hallidie Building (1918) in San Francisco. The first multistory structure with a full glass curtain wall was the A. O. Smith Research Building (1928) in Milwaukee by William Holabird and John Root; in it the glass was held by aluminum frames, an early use of this metal in buildings. But these were rare examples, and it was not until the development of

> air conditioning, fluorescent lighting, and synthetic rubber sealants after 1945 that the glass prism could be realized.
>
> The paradigm of the glass tower was defined by the United Nations Secretariat Building (1949) in New York City; Wallace Harrison was the executive architect, but Le Corbusier also played a major role in the design. The UN building, which featured a Weathermaster air-conditioning system and green-tinted glass walls, helped set the standard for tall buildings around the world. Several other influential buildings—such as Mies van der Rohe's 26-story 860–880 Lake Shore Drive Apartments (1951) in Chicago and Skidmore, Owings & Merrill's 21-story Lever House (1952) in New York City—helped to further establish the technology of curtain walls.

found after the exposition closed, and the building was demolished in 1910.

While these prodigious structures were the centre of attention, a new and more significant technology was developing: the steel-frame high-rise building. It began in Chicago, a city whose central business district was growing rapidly. The pressure of land values in the early 1880s led owners to demand taller buildings. The architect-engineer William Le Baron Jenney responded to this challenge with the 10-story Home Insurance Company Building (1885), which had a nearly completely all-metal structure. The frame consisted of cast-iron columns supporting wrought-iron beams, together with two floors of rolled-steel beams that were substituted during construction; this was the first large-scale use of steel in a building. The metal framing was completely encased in brick or clay-tile cladding for fire protection, since iron

Eiffel Tower, Paris. © Corbis

and steel begin to lose strength if they are heated above about 750°F (400°C). Jenney's Manhattan Building (1891) had the first vertical truss bracing to resist wind forces; rigid frame or portal wind bracing was first used in the neighbouring Old Colony Building (1893) by the architects William Holabird and Martin Roche. The all-steel frame finally appeared in Jenney's Ludington Building (1891) and the Fair Store (1892).

The foundations of these high-rise buildings posed a major problem, given the soft clay soil of central Chicago. Traditional spread footings, which dated back to the Egyptians, proved to be inadequate to resist settlement due to the heavy loads of the many floors, and timber piles (a Roman invention) were driven down to bedrock. For the 13-story Stock Exchange Building (1892), the engineer Dankmar Adler employed the caisson foundation used in bridge construction. A cylindrical shaft braced with board sheathing was hand-dug to bedrock and filled with concrete to create a solid pier to receive the heavy loads of the steel columns.

By 1895 a mature high-rise building technology had been developed: the frame of rolled steel I beams with bolted or riveted connections, diagonal or portal wind bracing, clay-tile fireproofing, and caisson foundations. The electric-powered elevator provided vertical transportation, but other environmental technologies were still fairly simple. Interior lighting was still largely from daylight, although supplemented by electric light. There was steam heating but no cooling, and ventilation was dependent on operating windows; thus these buildings needed narrow floor spaces to give adequate access to light and air. Of equal importance in high-rise construction was the introduction of the internal-combustion engine (which had been invented by Nikolaus Otto in 1876) at the

WILLIAM LE BARON JENNEY

American civil engineer and architect William Le Baron Jenney was a figure of primary importance in the development of the skyscraper. Jenney was born on Sept. 25, 1832, in Fairhaven, Mass. After studying architecture in Paris (1859–61), he served in the American Civil War (1861–65) as an engineering officer. Having left the Federal army with the rank of major, he practiced engineering and architecture in Chicago (1868–1905) and taught architecture at the University of Michigan, Ann Arbor (1876–80).

In Jenney's design for the Leiter Building, Chicago (1879; enlarged 1888; later demolished), he made a tentative approach to skeleton construction, and the facade was prophetic of the glass curtain wall that became common in the 20th century. Among his other buildings in Chicago are the Manhattan Building (1889–90), said to be the first 16-story structure in the world and the first in which wind bracing was a principal aspect of the design; the Ludington Building (1891); the Fair Store (1891–92; later remodelled as the Loop store of Montgomery Ward); and the second Leiter Building (1889–90), which became Sears, Roebuck and Co.'s Loop store.

Jenney designed the Home Insurance Company Building, Chicago (1884–85; enlarged 1891; demolished 1931), generally considered to be the world's first tall building supported by an internal frame, or skeleton, of iron and steel rather than by load-bearing walls and the first to incorporate steel as a structural material. The Home Insurance Company Building also set the pace for the Chicago School, many of whose chief exponents—including Louis Sullivan, Daniel Burnham, John Root, and William Holabird—served at one time in Jenney's office. Jenney died on June 15, 1907, in Los Angeles.

building site; it replaced the horse and human muscle power for the heaviest tasks of lifting.

THE ELEVATOR

By opening the way to the use of high-rise buildings, the elevator played a decisive role in creating the characteristic urban geography of many modern cities, especially in the United States, and promises to fill an indispensable role in future city development.

The practice of lifting loads by mechanical means during building operations goes back at least to Roman times. The Roman architect-engineer Vitruvius in the 1st century BCE described lifting platforms that used pulleys and capstans, or windlasses, operated by human, animal, or water power. Steam power was applied to such devices in England by 1800. In the early 19th century a hydraulic lift was introduced, in which the platform was attached to a plunger in a cylinder sunk in the ground below the shaft to a depth equal to the shaft's height. Pressure was applied to the fluid in the cylinder by a steam pump. Later a combination of sheaves was used to multiply the car's motion and reduce the depth of the plunger. All these devices employed counterweights to balance the weight of the car, requiring only enough power to raise the load.

Prior to the mid-1850s, these principles were primarily applied to freight hoists. The poor reliability of the ropes (generally hemp) used at that time made such lifting platforms unsatisfactory for passenger use. When an American, Elisha Graves Otis, introduced a safety device in 1853, he made the passenger elevator possible. Otis's device, demonstrated at the Crystal Palace Exposition in New York, incorporated a clamping arrangement that gripped the guide rails on which the car moved when tension was released from the hoist rope. The first passenger elevator was put into service in the Haughwout Department Store in New York City in 1857; driven by steam power, it climbed five stories in less than a minute and was a pronounced success.

Improved versions of the steam-driven elevator appeared in the next three decades, but no significant advance took place until the introduction of the electric motor for elevator operation in the mid-1880s and the first commercial installation of an electric passenger elevator in 1889. This installation, in the Demarest Building in New York City, utilized an electric motor to drive a winding drum in the building's basement. The introduction of electricity led to two further advances: in 1894 push-button controls were introduced, and in 1895 a hoisting apparatus was demonstrated in England that applied the power to the sheave (pulley) at the top of the shaft; the weights of the car and counterweight sufficed to guarantee traction. By removing the limitations imposed by the winding drum, the traction-drive mechanism made possible taller shafts and greater speeds. In 1904 a "gearless" feature was added by attaching the drive sheave directly to the armature of the electric motor, making speed virtually unlimited.

With the safety, speed, and height problems overcome, attention was turned to convenience and economy. In 1915 so-called automatic leveling was introduced in the form of automatic controls at each floor that took over when the operator shut off his manual control within a certain distance from the floor level and guided the car to a precisely positioned stop. Power control of doors was added. With increased building heights, elevator speeds increased to 1,200 feet (365 m) per minute in such express installations as those for the upper levels of the Empire State Building (1931) and reached 1,800 feet (549 m) per minute in the John Hancock Center, Chicago, in 1970.

Automatic operation, widely popular in hospitals and apartment buildings because of its economy, was improved by the introduction of collective operation, by which an elevator or group of elevators answered calls in sequence from top to bottom or vice versa. The basic safety feature of all elevator installations was the hoistway door interlock that required the outer (shaft) door to be closed and locked before the car could move. By 1950 automatic group-supervisory systems were in service, eliminating the need for elevator operators and starters.

HEATING, VENTILATION, AND AIR CONDITIONING

Heating, ventilation, and air conditioning, or HVAC, is an engineering discipline that is sometimes called climate control or environmental control. For modern buildings, systems are designed that control atmosphere and temperature year-round, in all weather conditions. But the concept of overall climate control is a recent development. Before the early 20th century, when it became possible to ventilate and cool buildings mechanically, climate control was limited mainly to the provision of heat during cold weather.

In most Roman buildings, a central open fire was the major source of heat—as well as annoying smoke—although the use of charcoal braziers made some improvement. A major innovation was the development of hypocaust, or indirect radiant, heating, by conducting heated air through flues in floors and walls. The heated masonry radiated a pleasantly uniform warmth, and smoke was eliminated from occupied spaces; the same method was used to heat water for baths. The Basilica of Constantine at Trier has a well-preserved example of hypocaust heating, where the stone slabs of the floor are supported on short brick columns, creating a continuous heating plenum beneath it.

Hypocaust heating disappeared with the Roman Empire, but a new development in interior heating appeared in western Europe at the beginning of the 12th century: the masonry fireplace and chimney began to replace the central open fire. The large roof openings over central fires let in wind and rain, so each house had only one and larger buildings had as few as possible. Therefore, heated rooms tended to be large and semipublic, where many persons could share the fire's warmth. The roof opening did not effectively remove all the smoke, some of which remained to plague the room's occupants. The chimney, on the other hand, did not let in much air or water and could remove most of the smoke. Although much of the heat went up the flue, it was still a great improvement, and, most significantly, it could be used to heat both small and large rooms and multistory buildings as well. Houses, particularly large ones, were broken up into smaller, more private spaces each heated by its own fireplace, a change that decisively altered the communal lifestyle of early medieval times.

During the Renaissance, the efficiency of interior heating was improved by the introduction of cast-iron and clay-tile stoves, which were placed in a free-standing position in the room. The radiant heat they produced was uniformly distributed in the space, and they lent themselves to the burning of coal—a new fuel that was rapidly replacing wood in western Europe.

Environmental control technologies began to develop dramatically with the Industrial Revolution. The stove and fireplace continued as the major sources of space heating throughout this period, but the development of the steam engine and its associated boilers led to a new technology in the form of steam heating. James Watt heated his own office with steam running through pipes as early as 1784. During the 19th century, systems of steam and later hot-water heating were gradually developed. These used coal-fired central boilers connected to networks of pipes that distributed the heated fluid to cast-iron radiators and returned it to the boiler for reheating. Steam heat was a major improvement over stoves and fireplaces because all combustion products were eliminated from occupied spaces, but heat sources were still localized at the radiators.

Steam and hot-water heating systems of the late 19th century provided a reasonable means for winter heating, but no practical methods existed for artificial cooling, ventilating, or humidity control. In the forced-air system of heating, air replaced steam or water as the fluid medium of heat transfer, but this was dependent on the development of powered fans to move the air. Large, crude fans for industrial applications in the ventilation of ships and mines had appeared by the 1860s, and the Johns Hopkins Hospital in Baltimore had a successful steam-powered forced-air system installed in 1873, but the widespread application of this system to buildings only followed the development of electric-powered fans in the 1890s.

Important innovations in cooling technology followed. The development of refrigeration machines for food storage played a role, but the key element was Willis Carrier's 1906 patent that solved the problem of humidity removal by condensing the water vapour on droplets of cold water sprayed into an airstream. Starting with humidity control in tobacco and textile factories, Carrier slowly developed his system of "man-made weather," finally applying it together with heating, cooling, and control devices as a complete system in Graumann's Metropolitan Theater, Los Angeles, in 1922. The first office building air-conditioned by Carrier was the 21-story Milam Building (1928) in San Antonio, Texas. It had a central refrigeration plant in the basement that supplied cold water to small air-handling units on every other floor. These units supplied conditioned air to each office space through ducts in the ceiling; the air was returned through grills in doors to the corridors and then back to the air-handling units. Carrier adopted a somewhat different system for the 32-story Philadelphia Savings Fund Society Building (1932). The central air-handling units were placed with the refrigeration plant on the 20th floor, and conditioned air was distributed through vertical ducts to the occupied floors and horizontally to each room and returned through the corridors to vertical exhaust ducts that carried it back to the central plant. Both systems of air handling, local and central, are still used in high-rise buildings.

Carrier's "Weathermaster" system was energy-intensive, appropriate to the declining energy costs of the post-World War II era, and it was adopted for most of the all-glass skyscrapers that followed in the next 25 years. In the 1960s the so-called dual-duct system appeared; both warm and cold air were centrally supplied to every part of the building and combined in mixing boxes to provide the appropriate atmosphere. The dual-duct system also consumed much energy. Thus, when energy prices began to rise in the 1970s, both the dual-duct system and the Weathermaster system were supplanted by the variable air volume (VAV) system, which supplies conditioned air at a single temperature, the volume varying according to the heat loss or gain in the occupied spaces. The VAV system requires much less energy and is still widely used.

Chapter 5: Medical Milestones

Nothing testifies more convincingly to human genius than the ability to understand the processes of the human body and to influence those processes in such a manner as to save lives, maintain good health, and fight off illness. Very often, milestones in our growing understanding of health and disease—from the invention of a vaccine to the discovery of an antibiotic to the transplanting of a living heart to the cloning of a living mammal—are reached by the efforts of specific individuals. Following are descriptions of some of the most notable of those milestones and profiles of the people who made them possible.

THE SMALLPOX VACCINE

Smallpox is caused by infection with variola major, a virus of the family *Poxviridae*. The disease is thought by some scholars to have arisen among settled agricultural populations in Mesopotamia as early as the 5th millennium BCE and in the Nile River Valley in the 3rd millennium BCE. The Persian physician ar-Rāzī, c. 900 CE, known as Rhazes to his European translators, clearly described the symptoms of smallpox and distinguished it from measles. Around 570, Bishop Marius of Avicentum (near Lausanne, Switz.) had introduced the Latin term *variola* (meaning "pox" or "pustule") to describe diseases that may have included smallpox. The English term *pox* was used to describe various eruptive diseases, including smallpox, but it was not until the 16th century that variola became popularly known as the "small pox," to distinguish it from syphilis (the "Great Pox"). By that time smallpox seems to have succeeded plague as the most feared pestilence in Europe. A huge pandemic reached from Europe to the Middle East in 1614, and epidemics arose regularly in Europe throughout the 17th and 18th centuries. Introduced to the Americas by European conquerors and settlers, smallpox decimated native populations from the Plains Indians of Canada and the United States to the Aztecs of Mexico and the Araucanians of Chile. The Australian Aboriginals also suffered large losses from the disease in the 19th century.

Though medical science had not yet grasped the concept of infectious organisms such as viruses, it was understood that smallpox was somehow a contagious disease and that its victims had to be separated from the general populace. Little could be done for the victims. The general approach to treating this disease, according to the great 17th-century English physician Thomas Sydenham, included procedures such as bloodletting, induction of vomiting, and administration of enemas in order to "keep the inflammation of the blood within due bounds." In all parts of the world, smallpox outbreaks brought on desperate pleas by believers for divine assistance in lifting the scourge. Beyond that, the only hope lay in preventing outbreaks from occurring.

In southwestern Asia it had been known for centuries that a healthy person could be made immune to smallpox by being injected with pus taken from the sores of an infected person. Another technique, practiced in China, was to grind the scabs of a smallpox victim and blow the powder through a tube into the nose of a healthy person. People inoculated in this way would suffer a brief illness themselves and would be contagious for a period, and a few would contract a serious infection and die. However, the risk of dying was far less than in a smallpox epidemic (roughly 2 percent, compared with 20 to 30 percent), and the benefit of immunity was clear. In the early 18th

EDWARD JENNER AND JAMES PHIPPS

In 1796 Edward Jenner was a successful physician in Berkeley, Gloucestershire, Eng., where he had been born 47 years before. Smallpox was widespread at the time, and occasional outbreaks of special intensity resulted in a very high death rate. Jenner had been impressed by the fact that a person who had suffered an attack of cowpox—a relatively harmless disease that could be contracted from cattle—could not take the smallpox—i.e., could not become infected whether by accidental or intentional exposure to smallpox. Pondering this phenomenon, Jenner concluded that cowpox not only protected against smallpox but could be transmitted from one person to another as a deliberate mechanism of protection.

The story of the great breakthrough is well known. In May 1796 Jenner found a young dairymaid, Sarah Nelmes, who had fresh cowpox lesions on her hand. On May 14, using matter from Sarah's lesions, he inoculated an eight-year-old boy, James Phipps, who had never had smallpox. Phipps became slightly ill over the course of the next 9 days but was well on the 10th. On July 1 Jenner inoculated the boy again, this time with smallpox matter. No disease developed; protection was complete. In 1798 Jenner, having added further cases, published privately a slender book entitled *An Inquiry into the Causes and Effects of the Variolae Vaccinae*. Vaccination rapidly proved its value, however, and Jenner became intensely active promoting it. The procedure spread rapidly to America and the rest of Europe and soon was carried around the world. Jenner died in Berkeley on Jan. 26, 1823.

century several European doctors and, most prominently, Lady Mary Wortley Montagu, the wife of Britain's ambassador to Turkey, began to publicize the value of inoculation (or variolation, as it came to be known), and the practice was soon adopted by royalty and people of means in Europe and America. However, the procedure was expensive and difficult, and it had little influence beyond the well-to-do classes and certain military forces.

A much safer procedure was developed by Edward Jenner, a physician in Gloucestershire, Eng. In 1796 Jenner deliberately infected a small boy with *variolae vaccinae*, or cowpox, a bovine version of smallpox. The boy suffered only a mild noncontagious reaction and then showed no reaction to a subsequent inoculation with *variola major*. The superiority of vaccination, as this technique came to be known, was immediately recognized. During the 19th century, vaccination programs, many of them compulsory, were instituted in many countries. At first vaccine was obtained directly from vaccinated persons, but soon it was being commercially harvested from pustules grown on the skin of inoculated calves. Later in the century it became apparent that the cowpox virus had been supplanted in vaccines by a different strain. It is still not certain whether the new virus, called *vaccinia*, was a mutation of cowpox virus or a completely separate strain, but it remains the virus used for vaccine production to this day.

By the beginning of the 20th century, smallpox was no longer endemic in several countries of continental Europe. Endemic smallpox was eradicated from the United Kingdom in 1934, the U.S.S.R. in 1936, Canada in 1946, the United States in 1949, Japan in 1951, and China in 1961. Still, in an age of global travel only an international effort could completely eradicate the disease. In 1967 the World Health Organization (WHO) began to vaccinate entire populations around every reported outbreak of smallpox. The disease was no longer endemic in India by 1975 and in Ethiopia by 1976. The last endemic case of smallpox (actually an infection of *variola minor*) was recorded in Somalia in 1977. No cases were reported from 1977 to 1980, with the exception of two cases in England in 1978 whose source was in a laboratory, and in 1980 the disease was declared exterminated.

GENERAL ANESTHESIA

The most famous contribution by the United States to medical progress during the 19th century was undoubtedly the introduction of general anesthesia, a procedure that not only liberated the patient from the fearful pain of surgery but also enabled the surgeon to perform more extensive operations. The discovery was marred by controversy, with several claimants for priority. There is little doubt, however, that it was dental surgeon William Thomas Morton who, on Oct. 16, 1846, at Massachusetts General Hospital, in Boston, first demonstrated before a gathering of physicians the use of ether as a general anesthetic.

The use of anesthetic gases in surgery was first proposed by British chemist Sir Humphry Davy in 1798, following his observation that inhalation of nitrous oxide, or "laughing gas," relieved pain. In the 1840s in New York City, Gardner Colton learned while studying medicine that the inhalation of nitrous oxide produced exhilaration. After a public demonstration of its effects proved to be a financial success, Colton began a lecture tour of other cities. On Dec. 10, 1844, at a Hartford, Conn., demonstration, Horace Wells, a dentist, asked Colton to extract one of his teeth while he was under the effects of the gas. Wells had earlier noted the pain-killing properties of nitrous oxide during a laughing-gas road show, and he thereafter used it in performing painless dental operations. He was allowed to demonstrate the method at the Massachusetts

General Hospital in January 1845, but when the patient proved unresponsive to the gas, Wells was exposed to ridicule.

Morton, Wells' former dental partner, was present at this unsuccessful demonstration. Determined to find a more reliable pain-killing chemical, Morton consulted his former teacher, Boston chemist Charles Jackson, who had learned from Wells that ether could be applied externally as a local anesthetic. The two discussed the use of ether, and Morton first used it in extraction of a tooth on Sept. 30, 1846. On October 16 he successfully demonstrated its use publicly, administering ether to a patient undergoing a tumour operation in the same theatre where Wells had failed nearly two years earlier.

Following Morton's successful demonstration, Wells began extensive self-experimentation with nitrous oxide, ether, chloroform, and other chemicals to ascertain their comparative anesthetic properties. His personality radically altered by frequent inhalation of chemical vapours, he took his own life in 1848 in a jail cell in New York City. Meanwhile, Morton attempted to obtain exclusive rights to the use of ether anesthesia. He spent the remainder of his life engaged in a costly contention with Jackson, who claimed priority in the discovery and denounced Morton as a swindler and a forger. In 1873 Jackson was admitted to an asylum for the mentally ill, where he died in 1880.

By some accounts credit for having been the first to use ether as an anesthetic in surgery goes to the rural Georgia physician Crawford Long. Long observed that persons injured in "ether frolics" (self-induced ether intoxication) seemed to suffer no pain, and in 1842 he painlessly removed a tumour from the neck of a patient to whom he had administered ether. He continued to use ether in other cases but did not publish any report of its use until 1849, well after Morton's public demonstration.

Regardless of controversy, the news of general anesthesia quickly reached Europe, and the practice soon became prevalent in surgery. At Edinburgh, the professor of midwifery, James Young Simpson, had been experimenting upon himself and his assistants, inhaling various vapours with the object of discovering an effective anesthetic. After news of the use of ether in surgery reached Scotland, Simpson tried it in obstetrics in January 1847. Later that year he substituted chloroform with complete success and published his classic *Account of a New Anaesthetic Agent*. Simpson persisted in the use of chloroform for relief of labour pains, against opposition of obstetricians and the clergy. He was appointed one of the queen's physicians for Scotland.

PASTEURIZATION

Pasteurization, the heat-treatment process that destroys pathogenic microorganisms in certain foods and beverages, is named for the French chemist and microbiologist Louis Pasteur, who in the 1860s arrived at the great invention indirectly, through the study of wine and beer. In 1854 Pasteur had been appointed professor of chemistry and dean of the science faculty at the University of Lille. While working at Lille, he was asked to help solve problems related to alcohol production at a local distillery, and thus he began a series of studies on alcoholic fermentation. His work on these problems led to his involvement in tackling a variety of other practical and economic problems involving fermentation. His efforts proved successful in unraveling most of these problems, and new theoretical implications emerged from his work. Pasteur investigated a broad range of aspects of fermentation, including the production of compounds such as lactic acid that are responsible for the souring of milk. He also studied butyric acid fermentation.

In 1857 Pasteur left Lille and returned to Paris, having been appointed manager and director of scientific studies at the École Normale Supérieure. That same year he presented experimental evidence for the participation of living organisms in all fermentative processes and showed that a specific organism was associated with each particular fermentation. This evidence gave rise to the germ theory of fermentation.

The realization that specific organisms were involved in fermentation was further supported by Pasteur's studies of butyric acid fermentation. These studies led Pasteur to the unexpected discovery that the fermentation process could be arrested by passing air (that is, oxygen) through the fermenting fluid, a process known today as the Pasteur effect. He concluded that this was due to the presence of a life-form that could function only in the absence of oxygen. This led to his introduction of the terms *aerobic* and *anaerobic* to designate organisms that live in the presence or absence of oxygen, respectively. He further proposed that the phenomena occurring during putrefaction were due to specific germs that function under anaerobic conditions.

Pasteur readily applied his knowledge of microbes and fermentation to the wine and beer industries in France, effectively saving the industries from collapse due to problems associated with production and with contamination that occurred during export. In 1863, at the request of the emperor of France, Napoleon III, Pasteur studied wine contamination and showed it to be caused by microbes. To prevent contamination, Pasteur used a simple procedure: he heated the wine to 120–140°F (50–60°C), a process now known universally as pasteurization.

Today, pasteurization is seldom used for wines that benefit from aging, since it kills the organisms that contribute to the aging process, but it is applied to many foods and beverages, particularly milk. Pasteurization is most important in all dairy processing. It is the biological safeguard that ensures that all potential pathogens are destroyed. Extensive studies have determined that heating milk to 145°F (63°C) for 30 minutes or 161°F (72°C) for 15 seconds kills the most resistant harmful bacteria. In actual practice these temperatures and times are exceeded, thereby not only ensuring safety but also extending shelf life.

Pasteurized milk is not sterile and is expected to contain small numbers of harmless bacteria. Therefore, the milk must be immediately cooled to below 40°F (4.4°C) and protected from any outside contamination. The shelf life for high-quality pasteurized milk is about 14 days when properly refrigerated.

X-RAY IMAGING

Perhaps the greatest modern anatomic diagnostic tool is the X-ray, serendipitously discovered by the German physicist Wilhelm Conrad Röntgen in his laboratory in the University of Würzburg on Nov. 8, 1895. While investigating the effects of electron beams (then called cathode rays) in electrical discharges through low-pressure gases, Röntgen uncovered a startling effect—namely, that a screen coated with a fluorescent material placed outside a discharge tube would glow even when it was shielded from the direct visible and ultraviolet light of the gaseous discharge. He deduced that an invisible radiation from the tube passed through the air and caused the screen to fluoresce. Röntgen was able to show that the radiation responsible for the fluorescence originated from the point where the electron beam struck the glass wall of the discharge tube. Opaque objects placed between the tube and the screen proved to be transparent to the new form of radiation; Röntgen dramatically demonstrated this by

ULTRASOUND

Much medical diagnostic imaging is carried out with X-rays but this type of radiation is highly ionizing—that is, X-rays are readily capable of destroying molecular bonds in the body tissue through which they pass. One of the important advantages of ultrasound (that is, high-frequency sound waves that are above the range of human hearing) is that it is a mechanical vibration and is therefore a nonionizing form of energy. Thus, it is usable in many sensitive circumstances where X-rays might be damaging. Also, the resolution of X-rays is limited owing to their great penetrating ability and the slight differences between soft tissues. Ultrasound, on the other hand, gives good contrast between various types of soft tissue.

Ultrasonic scanning in medical diagnosis uses the same principle as sonar. Pulses of high-frequency ultrasound, generally above 1 megahertz, are created by a piezoelectric transducer and directed into the body. As the ultrasound traverses various internal organs, it encounters changes in acoustic impedance, which cause reflections. The amount and time delay of the various reflections can be analyzed to obtain information regarding the internal organs and create an image.

Because it is nonionizing, ultrasound has become one of the staples of obstetric diagnosis. During the process of drawing amniotic fluid in testing for birth defects, ultrasonic imaging is used to guide the needle and thus avoid damage to the fetus or surrounding tissue. Ultrasonic imaging of the fetus can be used to determine the date of conception, to identify multiple births, and to diagnose abnormalities in the development of the fetus. Ultrasound is also used to provide images of the heart, liver, kidneys, gallbladder, breast, eye, and major blood vessels.

producing a photographic image of the bones of the human hand.

Early on, use was made of three of the properties of X-rays—their ability to penetrate the tissues, their photographic effect, and their ability to cause certain substances to fluoresce. In penetrating the tissues, the radiation is absorbed differentially, depending on the densities of the tissues being penetrated. The radiation emerging from tissues thus produces on a photographic film or a fluorescent screen an image of the structures of differing densities within the body. The limiting factor in this method of diagnosis is the similarity between the densities of adjacent soft tissues within the body, with a resultant failure to produce a notable contrast between the images of adjacent structures or organs.

During the first two decades following their discovery, X-rays were used largely for the diagnosis and control of treatment of fractures and for the localization of foreign bodies, such as bullets, during World War I. The physicians using these methods introduced artificial contrast agents, such as a paste consisting of barium sulfate, which is inert and nontoxic when taken by mouth. When a contrast agent is taken by mouth or introduced by enema, the various parts of the alimentary tract can be demonstrated and examined. Refinements of this technique continue to the present day, and radiological examination of the alimentary tract is a precise aid to diagnosis. Eventually a number of other contrast media were produced that could be injected into blood vessels. The media could thus be used either to demonstrate those vessels (whether arteries or veins) or, after their selective concentration and excretion by the kidneys, to show the urinary tract.

A new form of X-ray imaging, computerized axial tomography (CAT scanning), was devised by Godfrey Hounsfield of Great Britain and Allan Cormack of the

United States during the 1970s. This method measures the attenuation of X-rays entering the body from many different angles. From these measurements a computer reconstructs the organ under study in a series of cross sections or planes. The technique allows soft tissues such as the liver and kidney to be clearly differentiated in the images reconstructed by the computer. This procedure adds enormously to the diagnostic information that can be provided by conventional X-rays. CAT scanners are now in use throughout the world.

INSULIN

The vast majority of hormones were identified, had their biological activity defined, and were synthesized in the first half of the 20th century. Illnesses relating to their excess or deficiency were also beginning to be understood at that time. Hormones—produced in specific organs, released into the circulation, and carried to other organs—significantly affect metabolism and homeostasis. Some examples of hormones are insulin (from the pancreas), epinephrine (or adrenaline; from the adrenal medulla), thyroxine (from the thyroid gland), cortisol (from the adrenal cortex), estrogen (from the ovaries), and testosterone (from the testes). As a result of discovering these hormones and their mechanisms of action in the body, it became possible to treat illnesses of deficiency or excess effectively. The discovery and use of insulin to treat diabetes is an example of these developments.

In 1869 Paul Langerhans, a medical student in Germany, was studying the histology of the pancreas. He noted that this organ has two distinct types of cells—acinar cells, now known to secrete digestive enzymes, and islet cells (now called islets of Langerhans). The function of islet cells was suggested in 1889 when German physiologist and pathologist Oskar Minkowski and German physician Joseph von Mering showed that removing the pancreas from a dog caused the animal to exhibit a disorder quite similar to human diabetes mellitus (elevated blood glucose and metabolic changes). After this discovery, a number of scientists in various parts of the world attempted to extract the active substance from the pancreas so that it could be used to treat diabetes. It is now known that these attempts were largely unsuccessful because the digestive enzymes present in the acinar cells metabolized the insulin from the islet cells when the pancreas was disrupted.

In 1921 Fredrick Banting, a young Canadian surgeon in Toronto, convinced a physiology professor to allow him use of a laboratory to search for the active substance from the pancreas. Banting guessed correctly that the islet cells secreted insulin, which was destroyed by enzymes from the acinar cells. By this time Banting had enlisted the support of Charles Best, a fourth-year medical student. Together they tied off the pancreatic ducts through which acinar cells release the digestive enzymes. This caused the acinar cells to die. Subsequently, the remainder of the pancreas was homogenized and extracted with ethyl alcohol and acid. The extract thus obtained decreased blood glucose levels in dogs with a form of diabetes. Shortly thereafter, in 1922, a 14-year-old boy with severe diabetes was the first human to be treated successfully with the pancreatic extracts.

After this success other scientists became involved in the quest to develop large quantities of purified insulin extracts. Eventually, extracts from pig and cow pancreases created a sufficient and reliable supply of insulin. For the next 50 years most of the insulin used to treat diabetes was extracted from porcine and bovine sources. There are only slight differences in chemical structure between bovine, porcine, and human insulin, and their hormonal activities are essentially equivalent. Today, as a result of recombinant DNA technology, most of the insulin used in therapy is synthesized by pharmaceutical companies and is identical to human insulin.

ANTIBIOTICS

Antibiotics are chemical substances produced by living organisms, generally microorganisms, that are detrimental to other microorganisms. Antibiotics are released naturally into the soil by bacteria and fungi, but they did not come into worldwide prominence until the introduction of penicillin in 1941. Since then they have revolutionized the treatment of bacterial infections.

In 1928 Scottish bacteriologist Alexander Fleming noticed that colonies of the bacterium *Staphylococcus aureus* growing on a germ culture medium failed to grow in those areas that had been accidentally contaminated by the green mold *Penicillium notatum*. In spite of his conviction that penicillin was a potent antibacterial agent, Fleming was unable to carry his work to fruition, mainly because biochemists at the time were unable to isolate it in sufficient quantities or in a sufficiently pure form to allow its use on patients. A decade later British biochemist Ernst Chain, Australian pathologist Howard Florey, and others isolated the ingredient responsible, penicillin, in a form that was fairly pure (by standards then current) and showed that it was highly effective against many serious bacterial infections. In 1941 an injectable form of the drug was available for therapeutic use. By then World War II had begun, and techniques to facilitate commercial production were developed in the United States. By 1944 adequate amounts were available to meet the extraordinary needs of wartime.

Toward the end of the 1950s scientists added various chemical groupings to the core of the penicillin molecule to generate semisynthetic versions. A range of penicillins is thus now available to treat diseases caused by such bacteria as staphylococci, streptococci, pneumococci, gonococci, and the spirochaetes of syphilis.

Conspicuously unaffected by penicillin is the tubercle bacillus, *Mycobacterium tuberculosis*, but this organism proved to be highly sensitive to streptomycin, isolated from *Streptomyces griseus* in 1943. As well as being dramatically effective against tuberculosis, streptomycin also vanquishes many other bacteria, including the typhoid fever bacillus. Two other early discoveries were gramicidin and tyrocidin, made by bacteria of the genus *Bacillus*. Discovered in 1939 by René Dubos, they have proved to be valuable in treating surface infections but are too toxic for internal use. Other, more recently isolated, antibiotic drugs are the cephalosporins. Related to penicillins, they are produced by the mold *Cephalosporium acremonium*. A class of antibodies first developed in the 1960s, called quinolones, interrupt the replication of DNA (a crucial step in bacterial reproduction) and have proved useful in treating urinary-tract infections, infectious diarrhea, and various other infections involving such elements as bones and white blood cells. Many bacterial and, to a lesser extent, fungal infections can be treated by antibiotics; such drugs cannot treat viral infections, however.

A problem that has plagued antibiotic therapy from the earliest days is the resistance that bacteria can develop to the drugs. An antibiotic may kill virtually all the bacteria causing a disease in a patient, but a few bacteria that are genetically less vulnerable to the effects of the drug may survive. These go on to reproduce or to transfer their resistance to others of their species through processes of gene exchange. With their more vulnerable competitors wiped out or reduced in numbers by antibiotics, these resistant strains proliferate; the end result is bacterial infections in humans that are untreatable by one or even several of the antibiotics customarily effective in such cases. The indiscriminate and inexact use of antibiotics encourages the spread of such bacterial resistance.

BLOOD TRANSFUSION

In 1616 English physician William Harvey announced his observations on the circulation of the blood, and in 1628 he published his famous monograph titled *Exercitatio Anatomica de Motu Cordis et Sanguinis in Animalibus* (*The Anatomical Exercises Concerning the Motion of the Heart and Blood in Animals*). His discovery, that blood circulates around the body in a closed system, was an essential prerequisite of the concept of transfusing blood from one animal to another of the same or different species. In England, experiments on the transfusion of blood were pioneered in dogs in 1665 by physician Richard Lower. In November 1667 Lower transfused the blood of a lamb into a man. Meanwhile, in France, Jean-Baptiste Denis, court physician to King Louis XIV, had also been transfusing lambs' blood into human subjects and described what is probably the first recorded account of the signs and symptoms of a hemolytic transfusion reaction. Denis was arrested after a fatality, and the procedure of transfusing the blood of other animals into humans was prohibited, by an act of the Chamber of Deputies in 1668, unless sanctioned by the Faculty of Medicine of Paris. Ten years later, in 1678, the British Parliament also prohibited transfusions. Little advance was made in the next 150 years.

In England in the 19th century, interest was reawakened by the activities of obstetrician James Blundell, whose humanitarian instincts had been aroused by the frequently fatal outcome of hemorrhage occurring after childbirth. He insisted that it was better to use human blood for transfusion in such cases.

In 1875 German physiologist Leonard Landois showed that, if the red blood cells of an animal belonging to one species are mixed with serum taken from an animal of another species, the red cells usually clump and sometimes burst—i.e., hemolyze. He attributed the appearance of black urine after transfusion of heterologous blood (blood from a different species) to the hemolysis of the incompatible red cells. Thus, the dangers of transfusing blood of another species to humans were established scientifically.

The human ABO blood groups were discovered by Austrian-born American biologist Karl Landsteiner in 1901. Landsteiner found that there are substances in the blood, antigens and antibodies, that induce clumping of red cells when red cells of one type are added to those of a second type. He recognized three groups—A, B, and O—based on their reactions to each other. A fourth group,

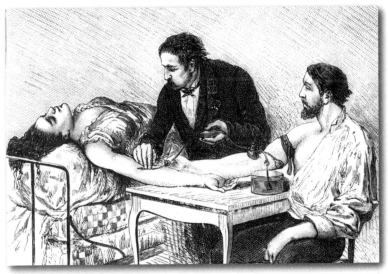

Depiction of an 1882 direct blood transfusion. © Photos.com/Jupiterimages

CHARLES DREW

Charles Richard Drew was an African American physician and surgeon who was an authority on the preservation of human blood for transfusion.

Drew was born on June 3, 1904, in Washington, D.C. He was educated at Amherst College (he graduated in 1926), McGill University, Montreal (1933), and Columbia University (1940). While Drew was earning his doctorate at Columbia University in the late 1930s, he conducted research into the properties and preservation of blood plasma. He soon developed efficient ways to process and store large quantities of blood plasma in "blood banks." As the leading authority in the field, he organized and directed the blood-plasma programs of the United States and Great Britain in the early years of World War II, while also agitating the authorities to stop excluding the blood of African Americans from plasma-supply networks.

Drew resigned his official posts in 1942 after the armed forces ruled that the blood of African Americans would be accepted but would have to be stored separately from that of whites. He then became a surgeon and professor of medicine at Freedmen's Hospital, Washington, D.C., and Howard University (1942–50). He was fatally injured in an automobile accident and died on April 1, 1950, near Burlington, N.C. Within a short time, rumours arose that the injured doctor had died after he had been refused treatment—including a blood transfusion—at a segregated hospital. Subsequent historical research, taking into account statements and records by Drew's family, his colleagues, and hospital staff, has shown the story to be a myth.

AB, was identified a year later by another research team. Red cells of the A group clump with donor blood of the B group; those of the B group clump with blood of the A group; those of the AB group clump with those of the A or the B group because AB cells contain both A and B antigens; and those of the O group do not generally clump with any group, because they do not contain either A or B antigens. The application of knowledge of the ABO system in blood transfusion practice is of enormous importance, since mistakes can have fatal consequences. In 1914 sodium citrate was added to freshly drawn blood to prevent clotting. Blood was occasionally transfused during World War I, but three-quarters of a pint was considered a large amount. These transfusions were given by directly linking the vein of a donor with that of the recipient. The continuous drip method, in which blood flows from a flask, was introduced by Hugh Marriott and Alan Kekwick at the Middlesex Hospital, London, in 1935.

The discovery of the Rh system by Landsteiner and Alexander Wiener in 1940 was made because they tested human red cells with antisera (animal or human serum containing antibodies specific for one or more antigens) developed in rabbits and guinea pigs by immunization of the animals with the red cells of the rhesus monkey *Macaca mulatta*. Other blood groups were identified later, such as Kell, Diego, Lutheran, Duffy, and Kidd. The remaining blood group systems were first described after antibodies were identified in patients. Frequently, such discoveries resulted from the search for the explanation of an unexpected unfavourable reaction in a recipient after a transfusion with formerly compatible blood. In such cases the antibodies in the recipient were produced against previously unidentified antigens in the donor's blood. In the case of the Rh system, for example, the presence of antibodies in the maternal serum directed against antigens present on the child's red cells can have serious consequences because of antigen-antibody reactions that produce erythroblastosis fetalis, or hemolytic disease of the newborn. Some of the other blood group systems—for example, the Kell and Kidd systems—were discovered because an infant was found to have erythroblastosis fetalis even though mother and child were compatible as far as the Rh system was concerned.

As blood transfusions increased in frequency and volume, blood banks were required. Although it took another world war before these were organized on a large scale, the first tentative steps were taken by Sergey Sergeyevich Yudin, of Moscow, who, in 1933, used cadaver blood, and by Bernard Fantus, of Chicago, who, four years later, used living donors as his source of supply. Saline solution, plasma, artificial plasma expanders, and other solutions are now also used in the appropriate circumstances.

POLIO VACCINE

Polio epidemics did not begin to occur until the latter part of the 19th century, but evidence indicates that it is an ancient disease. A well-known stele from the 18th dynasty of ancient Egypt (1570–1342 BCE) clearly depicts a priest with a telltale paralysis and withering of his lower right leg and foot. The mummy of the pharaoh Siptah from the late 19th dynasty (1342–1197 BCE) shows a similarly characteristic deformity of the left leg and foot. However, owing to the sporadic appearance of the infection, the absence of epidemics until relatively recent times, and the nonspecific nature and infrequency of the acute illness, there is hardly another recognizable trace of the disease until the 18th century. In 1789 a pediatrician in London, Michael Underwood, published the first clear description of paralytic disease of infants (as polio was then called) in a medical textbook. In the early 19th century, small groups of polio-afflicted patients began to be reported in the medical literature, but still only as sporadic cases.

It is an irony of medical history that the transformation of polio into an epidemic disease occurred only in those industrialized countries in North America and Europe that had experienced significant improvements in hygiene during the 19th and 20th centuries. This has led health experts to conjecture that the infection was common in earlier times but that people were exposed and infected (in typically unhygienic environments) at very young ages, when they were less likely to suffer permanent paralysis as an outcome. As hygiene improved, the certainty of young people of successive generations being exposed to the virus was gradually reduced; in this new situation it was not long before enough susceptible children and adults had accumulated to allow epidemics to break out.

The first epidemics appeared in the form of outbreaks of at least 14 cases near Oslo, Norway, in 1868, and of 13 cases in northern Sweden in 1881. About the same time the idea began to be suggested that the hitherto sporadic cases of infantile paralysis might be contagious. The next significant epidemic—10 times larger than previous outbreaks, with 132 recognized cases—erupted in the U.S. state of Vermont in 1894. During an epidemic of 1,031 cases in Sweden in 1905, Ivar Wickman recognized that patients with nonparalytic disease could spread the virus, and during an epidemic of 3,840 cases in 1911, Carl Kling and colleagues in Stockholm recovered the virus from healthy carriers as well as paralytic patients. In studying several fatal cases from the same outbreak, Kling found the virus in the victims' throats and also in tissues of their small intestines. During the second decade of the 20th century, it became apparent that far more people were being rendered immune to polio by previous asymptomatic infections than were being immunized by recovery from overt disease. By then polio was well on the way to becoming a widely feared periodic phenomenon. In the 1940s and early 1950s, western Europe and North America lived through summertime terrors brought about by nearly annual polio epidemics. At its peak incidence in the United States, in 1952, approximately 21,000 cases of paralytic polio (a rate of 13.6 cases per 100,000 population) were recorded. As outbreaks were concentrated in the summer and early autumn, children were kept away from swimming pools, movie theatres, and other crowded places where they might be exposed to the dreaded virus. Outbreaks were widely reported in the press, and polio victims encased in iron lungs were often displayed in public places such as department stores in order to encourage donations to efforts to research and combat the disease. In such an environment, it is not surprising that the announcement of an effective vaccine in 1955 was hailed as a mid-20th-century miracle.

The poliovirus itself was discovered in 1908 by a team led by Viennese immunologist and future Nobel Prize winner Karl Landsteiner. The existence of telltale antibodies specific to the virus circulating in the blood of infected persons was discovered only two years later. In 1931 two Australian researchers, Frank Macfarlane Burnet and Jean Macnamara, using immunologic techniques, were able to identify the different serotypes of the poliovirus. (Burnet was to receive a Nobel Prize in 1960.) In 1948 the team of John Enders, Thomas Weller, and Frederick Robbins, working at Harvard Medical School in Massachusetts, showed how the virus could be grown in large amounts in tissue culture (an advance for which they shared a Nobel Prize in 1954). From there it was only a short step to an announcement in 1953 by Jonas Salk at the University of Pittsburgh, Pennsylvania, that he had developed an effective killed-virus vaccine.

Salk's vaccine, known as the inactivated poliovirus vaccine (IPV), was put to a massive nationwide test in 1954–55. Called the Francis Field Trial after Thomas Francis Jr., a University of Michigan professor who directed it, the test involved 1.8 million children in the first, second, and third grades across the United States. The trial was declared a success on April 12, 1955, and over the next four years more than 450 million doses of the Salk vaccine were distributed. During that time the incidence of paralytic polio in the United States fell from 18 cases per 100,000 population to fewer than 2 per 100,000. In the years 1961–63, approval was given to a new vaccine developed by Albert Sabin at the University of Cincinnati, Ohio. The Sabin vaccine, using live but attenuated virus, could be given in drops through the mouth and therefore became known as the oral poliovirus vaccine (OPV). Soon it became the predominant vaccine used in the United States and most other countries. By the early 1970s the annual incidence of polio in the United States had declined a thousandfold from prevaccine levels, to an average of 12 cases a year.

THE BIRTH CONTROL PILL

The fact that conception was more likely to take place during certain phases of the menstrual cycle than others was suspected by classical authors. Adam Raciborski, a Paris physician, noted in 1843 that brides married soon after their menstruation often conceived in that cycle, while if the wedding occurred later in the cycle they commonly had another period before pregnancy occurred. Hermann Knaus in Austria (1929) and Kyusaku Ogino in

Japan (1930) independently and correctly concluded that ovulation occurs 14 days prior to the next menstruation. In 1964 an Australian medical team, John and Evelyn Billings, showed how women could monitor changes in their cervical mucus and learn to predict when ovulation would occur.

"The greatest invention some benefactor can give mankind," wrote Sigmund Freud in the early years of the 20th century, "is a form of contraception which does not induce neurosis." Many of the elements to meet the goal of a new, more acceptable form of contraception were present about the time of World War I, yet two generations were to reach maturity before those elements were exploited. The role of hormones from the ovary was understood early in the 20th century by Walter Heape and John Marshall. The first extract of estrogen was produced in 1913, and the pure compound was isolated by the Americans Willard Allen and Alan Doisy in 1923. At this time an Austrian physiologist, Ludwig Haberlandt, was carrying out experiments on rabbits to apply the new-found knowledge of hormones for contraceptive ends. By 1927 he was able to write, "It needs no amplification, of all methods available, hormonal sterilization based on biologic principles, if it can be applied unobjectionably in the human, is an ideal method for practical medicine and its future task of birth control." Hostile public attitudes made research on birth control virtually impossible, however.

Though the principle of hormonal contraception was understood in the 1920s, it was 30 years before the drive of American social reformer Margaret Sanger (then more than 70 years old) and the philanthropy of Katharine McCormick were to draw the first oral contraceptive preparations from somewhat reluctant scientists and physicians. Sanger, a trained nurse, had encountered miserable conditions in her work among the poor. She was inspired to take up her crusade when she attended a woman who was dying from a criminally induced abortion. In 1914 she started a magazine, *The Woman Rebel*, to challenge laws restricting the distribution of information on birth control. She was indicted and fled to Europe, but when she returned to stand trial in 1916 the charges against her were dropped. Later that year she opened a family planning clinic in Brownsville, Brooklyn, N.Y., but the police immediately closed it, and Sanger was arrested and convicted on charges of "maintaining a public nuisance." After many vicissitudes, a compromise was struck and family planning clinics were allowed in the United States on the condition that physicians be involved in prescribing contraceptives. In 1936 a New York court, in a case known as *United States v. One Package of Japanese Pessaries*, ruled that contraceptives could be sent through the post if they were to be intelligently employed by conscientious physicians for the purpose of saving life or promoting the well-being of their patients.

The first clinical report of the use of such preparations to suppress ovulation was published in 1956 by American endocrinologist Gregory Pincus and American gynecologist and researcher John Rock. In 1944 Pincus and Hudson Hoagland founded the Worcester Foundation for Experimental Biology, which became an important centre for the study of steroid hormones and mammalian reproduction. Sanger encouraged his work, and in 1951 Pincus and his collaborators began to work with synthesized hormones and the prevention of pregnancy. They found that inhibition of ovulation was an effective means of preventing pregnancy in laboratory animals and moved to perfect an oral contraceptive for women. Oral contraceptives were approved by the U.S. Food and Drug Administration in 1960, and marketing of the preparations in Britain began two years later.

There are many commercial preparations of oral contraceptives, but most of them contain a combination of an estrogen (usually ethinyl estradiol) and a progestin (commonly norethindrone). Oral contraceptives contain synthetic steroid hormones that suppress the release of follicle-stimulating hormone (FSH) and luteinizing hormone (LH) from the anterior lobe of the pituitary gland in the female body. FSH and LH normally stimulate the release of estrogen from the ovaries, which in turn stimulates ovulation—the release of a mature egg from the female ovary. However, when FSH and LH are suppressed, the chances of ovulation and therefore fertilization by a male sperm cell are significantly reduced. When oral contraceptives are used correctly, they are between 92 and 99 percent effective in preventing an unintended pregnancy.

In general, oral contraceptives are taken in a monthly regimen that parallels the menstrual cycle. Protection from pregnancy is often unreliable until the second or third drug cycle, and during this time certain side effects such as nausea, breast tenderness, or bleeding may be evident. More serious side effects, including blood clots and a rise in blood pressure, are possible, especially in women over 34 years of age. However, the incidence of side effects from oral contraceptives has been significantly reduced by decreasing the amounts of estrogen and progesterone in the preparations. Normal ovulation usually commences two to three months after the drug is stopped.

HEART TRANSPLANTATION

In 1967 surgery arrived at a climax that made the whole world aware of its medicosurgical responsibilities when the South African surgeon Christiaan Neethling Barnard transplanted the first human heart. Reaction, both medical and lay, contained more than an element of hysteria.

Yet research had been relentlessly leading up to just such an operation ever since Charles Guthrie and Alexis Carrel, at the University of Chicago, perfected the suturing of blood vessels in 1905 and then carried out experiments in the transplantation of many organs, including the heart.

The heart is a pump with a built-in power supply; it has a delicate regulatory mechanism that permits it to perform efficiently under a wide range of demands. During moments of fear, passion, or vigorous exercise, the heart rate increases greatly, and the contractions become more forceful, so that the pumping of the blood intrudes on the consciousness; this is experienced by the individual as palpitations. Cessation of the heartbeat has also been, throughout the ages, the cardinal sign of death. Thus, it is perhaps not so surprising that there was an intense public interest when the first attempts were made at transplanting a human heart. The objectives of heart transplantation, nevertheless, are the same as those of other organ grafts.

A group of American investigators perfected the technique of heart transplantation in the late 1950s. They showed that a transplanted dog's heart could provide the animal with a normal circulation until the heart was rejected. The features of rejection of the heart are similar to those of the kidney. The cells that produce immune reactions, the lymphocytes, migrate into the muscle cells of the heart, damage it, and also block the coronary arteries, depriving the heart of its own circulation. Some of the lymphocytes (i.e., B cells) also secrete antibodies that are toxic. In most experiments it was more difficult to prevent rejection of the heart than of the kidney. Despite this, rejection was prevented for long periods in animals. Based on this experimental work, the next logical step was to transplant a human heart into a patient dying of incurable heart disease. This step was taken in 1967 by the surgical team in Cape Town, S.Af.

THE ARTIFICIAL HEART

Any device that maintains blood circulation and oxygenation in the human body for varying periods of time is called an artificial heart. There are two main types: the heart-lung machine and the mechanical heart.

The heart-lung machine is a mechanical pump that maintains a patient's blood circulation and oxygenation

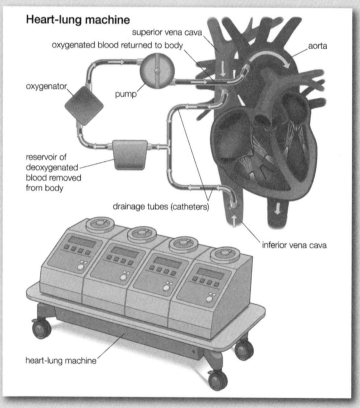

A heart-lung machine is connected to the heart by drainage tubes that divert blood from the venous system, directing it to an oxygenator. The oxygenator removes carbon dioxide and adds oxygen to the blood, which is then returned to the arterial system of the body. Encyclopædia Britannica, Inc.

during heart surgery by diverting blood from the venous system, directing it through tubing into an artificial lung (oxygenator), and returning it to the body. The first successful clinical use of a heart-lung machine was reported by American surgeon John H. Gibbon Jr. in 1953. During this operation for the surgical closure of an atrial septal defect, cardiopulmonary bypass was achieved by a machine equipped with an oxygenator developed by Gibbon and a roller pump developed in 1932 by American surgeon Michael E. DeBakey. Since then, heart-lung machines have been greatly improved with smaller and more-efficient oxygenators, allowing them to be used not only in adults but also in children and even newborn infants.

Mechanical hearts, which include total artificial hearts and ventricular assist devices (VADs), are machines that are capable of replacing or assisting the pumping action of the heart for prolonged periods without causing excessive damage to the blood components. Implantation of a total artificial heart requires removal of both of the patient's ventricles (lower chambers). However, with the use of a VAD to support either the right or the left ventricle, the entire heart remains in the body.

The first successful use of a mechanical heart in a human was performed by Michael E. DeBakey in 1966. After surgery to replace the patient's aorta and mitral valve, a left VAD was installed, making it possible to wean the patient from the heart-lung machine. After 10 days of pump flow from the VAD, the heart recovered, and the VAD was removed. During the 1970s synthetic materials were developed that greatly aided the development of permanent artificial hearts. One such device, designed by American physician Robert K. Jarvik, was surgically implanted into a patient by American surgeon William C. DeVries in 1982. The aluminum and plastic device, called the Jarvik-7 for its

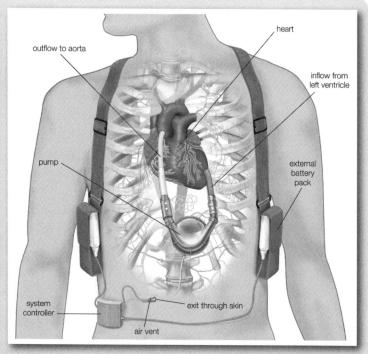

Ventricular assist device (VAD), a type of artificial heart designed to assist one of the ventricles (in this case the left) in pumping oxygenated blood through the aorta and to the body's tissues. The pump is placed inside the chest cavity, while the power source and system controller are carried on a harness outside the body. Encyclopædia Britannica, Inc.

inventor, replaced the patient's two ventricles. Two rubber diaphragms, designed to mimic the pumping action of the natural heart, were kept beating by an external compressor that was connected to the implant by hoses. This first recipient survived 112 days and died as a result of various physical complications caused by the implant. Subsequent patients fared little better or even worse, so that use of the Jarvik-7 was stopped. In 2001 a team of American surgeons implanted the first completely self-contained artificial heart, called the AbioCor artificial heart. The patient survived 151 days.

As a resident surgeon at Groote Schuur Hospital, Cape Town (1953–56), Christiaan Neethling Barnard was the first to show that intestinal atresia, a congenital gap in the small intestine, is caused by an insufficient blood supply to the fetus during pregnancy. This discovery led to the development of a surgical procedure to correct the formerly fatal defect. After completing doctoral studies at the University of Minnesota (1956–58), he returned to the hospital as senior cardiothoracic surgeon, introduced open-heart surgery to South Africa, developed a new design for artificial heart valves, and began extensive experimentation on heart transplantation in dogs. On Dec. 3, 1967, Barnard led a team of 20 surgeons in replacing the heart of Louis Washkansky, an incurably ill South African grocer, with a heart taken from a fatally injured accident victim. Although the transplant itself was successful, Washkansky died 18 days later from double pneumonia, contracted after destruction of his body's immunity mechanism by drugs administered to suppress rejection of the new heart as a foreign protein. Barnard's later transplant operations were increasingly successful; by the late 1970s a number of his patients had survived for several years.

In the years immediately following the first transplant, numerous heart allografts were performed at medical centres throughout the world. Unfortunately, many recipients succumbed to rejection of the transplanted organ. Furthermore, the heart is more sensitive to lack of blood than the kidneys are; it must be removed from the donor more quickly and can be preserved without damage for only a short period of time. Because of these difficulties—particularly the problem of rejection—the number of heart transplants performed worldwide dropped considerably after the initial excitement abated. Steady advances in detecting and treating rejection were made throughout the 1970s, however; and the introduction of the immunosuppressant cyclosporine in the 1980s brought even further improvements in the long-term survival rates for heart-graft recipients. Interest in the procedure revived, and many hundreds of heart transplants have now been performed. A number of patients have lived five or more years after the operation, and heart grafting has become an accepted therapy for otherwise incurable heart disease. Experimental artificial hearts have also been implanted, but these require a cumbersome external power supply and long-term survival rates are not known.

GENETIC ENGINEERING

The term *genetic engineering* initially meant any of a wide range of techniques for the modification or manipulation of organisms through the processes of heredity and reproduction. As such, the term embraced both artificial selection and all the interventions of biomedical techniques, among them artificial insemination, in vitro fertilization (e.g., "test-tube" babies), sperm banks, cloning, and gene manipulation. But today the term denotes the narrower field of recombinant DNA technology, or gene cloning, in which DNA molecules from two or more sources are combined either within cells or in vitro and are then inserted into host organisms in which they are able to propagate. Gene cloning is used to produce new genetic combinations that are of value to science, medicine, agriculture, or industry.

DNA is the carrier of genetic information; it achieves its effects by directing the synthesis of proteins. Most recombinant DNA technology involves the insertion of foreign genes into the plasmids of common laboratory strains of bacteria. Plasmids are small rings of DNA; they are not part of the bacterium's chromosome (the main repository of the organism's genetic information).

Nonetheless, they are capable of directing protein synthesis, and, like chromosomal DNA, they are reproduced and passed on to the bacterium's progeny. Thus, by incorporating foreign DNA (for example, a mammalian gene) into a bacterium, researchers can obtain an almost limitless number of copies of the inserted gene. Furthermore, if the inserted gene is operative (i.e., if it directs protein synthesis), the modified bacterium will produce the protein specified by the foreign DNA.

A key step in the development of genetic engineering was the discovery of restriction enzymes in 1968 by the Swiss microbiologist Werner Arber. However, type II restriction enzymes, which are essential to genetic engineering for their ability to cleave a specific site within the DNA (as opposed to type I restriction enzymes, which cleave DNA at random sites), were not identified until 1969, when the American molecular biologist Hamilton O. Smith purified this enzyme. Drawing on Smith's work, the American molecular biologist Daniel Nathans helped advance the technique of DNA recombination in 1970–71 and demonstrated that type II enzymes could be useful in genetic studies. Genetic engineering itself was pioneered in 1973 by the American biochemists Stanley N. Cohen and Herbert W. Boyer, who were among the first to cut DNA into fragments, rejoin different fragments, and insert the new genes into *E. coli* bacteria, which then reproduced.

Genetic engineering has advanced the understanding of many theoretical and practical aspects of gene function and organization. Through recombinant DNA techniques, bacteria have been created that are capable of synthesizing human insulin, human growth hormone, alpha interferon, a hepatitis B vaccine, and other medically useful substances. Plants may be genetically adjusted to enable them to fix nitrogen, and genetic diseases can possibly be corrected by replacing "bad" genes with "normal" ones. Nevertheless, special concern has been focused on such achievements for fear that they might result in the introduction of unfavourable and possibly dangerous traits into microorganisms that were previously free of them—e.g., resistance to antibiotics, production of toxins, or a tendency to cause disease.

The "new" microorganisms created by recombinant DNA research were deemed patentable in 1980, and in 1986 the U.S. Department of Agriculture approved the sale of the first living genetically altered organism—a virus, used as a pseudorabies vaccine, from which a single gene had been cut. Since then several hundred patents have been awarded for genetically altered bacteria and plants.

CLONING

Cloning is the process of generating a genetically identical copy of a cell or an organism. Cloning happens all the time in nature—for example, when a cell replicates itself asexually without any genetic alteration or recombination. Prokaryotic organisms (organisms lacking a cell nucleus), such as bacteria and yeasts, create genetically identical duplicates of themselves using binary fission or budding. In eukaryotic organisms (organisms possessing a cell nucleus) such as humans, all the cells that undergo mitosis, such as skin cells and cells lining the gastrointestinal tract, are clones; the only exceptions are gametes (eggs and sperm), which undergo meiosis and genetic recombination.

In biomedical research cloning is broadly defined to mean the duplication of any kind of biological material for scientific study, such as a piece of DNA or an individual cell. For example, segments of DNA are replicated exponentially by a process known as polymerase chain reaction, or PCR, a technique that is used widely in basic biological research. The type of cloning that is the focus of much ethical controversy involves the generation of cloned embryos, particularly those of humans, which are genetically identical to the organisms from which they are derived, and the subsequent use of these embryos for research, therapeutic, or reproductive purposes.

Early Cloning Experiments

Reproductive cloning was originally carried out by artificial "twinning," or embryo splitting, which was first performed on a salamander embryo in 1902 by German embryologist Hans Spemann. In 1928, Spemann, who was later awarded the Nobel Prize for Physiology or Medicine (1935) for his research on embryonic development, theorized about another cloning procedure known as nuclear transfer. This procedure was performed in 1952 by American scientists Robert W. Briggs and Thomas J. King, who used DNA from frog embryonic cells to generate cloned tadpoles. A decade later, British biologist John Gurdon successfully carried out nuclear transfer using DNA from adult frog cells.

Advancements in the field of molecular biology led to the development of techniques that allowed scientists to manipulate cells and to detect chemical markers that signal changes within cells. With the advent of recombinant DNA technology in the 1970s, it became possible for scientists to create transgenic clones—clones with genomes containing pieces of DNA from other organisms. Beginning in the 1980s mammals such as sheep were cloned from early and partially differentiated embryonic cells. In 1996 British developmental biologist Ian Wilmut generated a cloned sheep, named Dolly, by means of nuclear transfer involving an enucleated embryo and a differentiated cell nucleus. This technique, which was later refined and became known as somatic cell nuclear transfer (SCNT), represented an extraordinary advance in the science of cloning, because it resulted in the creation of a genetically identical clone of an already grown sheep. It also indicated that it was possible for the DNA in differentiated somatic (body) cells to revert to an undifferentiated embryonic stage, thereby reestablishing pluripotency—the potential of an embryonic cell to grow into any one of the numerous different types of mature body cells that make up a complete organism. The realization that the DNA of somatic cells could be reprogrammed to a pluripotent state significantly impacted research into therapeutic cloning and the development of stem cell therapies.

Soon after the generation of Dolly, a number of other animals were cloned by SCNT, including pigs, goats, rats, mice, dogs, horses, and mules. Despite these successes, the birth of a viable SCNT primate clone has not been achieved. In 2001 a team of scientists cloned a rhesus monkey through a process called embryonic cell nuclear transfer, which is similar to SCNT except that it uses DNA from an undifferentiated embryo. In 2007 macaque monkey embryos were cloned by SCNT; however, these clones lived only to the blastocyst stage of embryonic development. Likewise, SCNT has been carried out with very limited success in humans.

Reproductive Cloning

Reproductive cloning involves the implantation of a cloned embryo into a real or an artificial uterus. The embryo develops into a fetus that is then carried to term. Reproductive cloning experiments were performed for more than 40 years through the process of embryo splitting, in which a single early-stage two-cell embryo is manually divided into two individual cells and then grows as two identical embryos. Reproductive cloning techniques underwent significant change in the 1990s, following the birth of Dolly, who was generated through the process of SCNT. This process entails the removal of the entire nucleus from a somatic (body) cell of an organism, followed by insertion of the nucleus into an egg cell that has had its own nucleus removed (enucleation). Once the somatic nucleus is inside the egg, the egg is stimulated with a mild electrical current and begins dividing. Thus, a cloned embryo, essentially an embryo of an identical twin of the original organism, is created. The SCNT process has undergone significant refinement since the 1990s, and procedures have been developed to prevent damage to eggs during nuclear extraction and somatic cell nuclear insertion. For example, the use of polarized light to visualize an egg cell's nucleus facilitates the extraction of the nucleus from the egg, resulting in a healthy, viable egg and thereby increasing the success rate of SCNT.

Reproductive cloning using SCNT is considered very harmful since the fetuses of embryos cloned through SCNT rarely survive gestation and usually are born with birth defects. Wilmut's team of scientists needed 227 tries to create Dolly. Likewise, attempts to produce a macaque monkey clone in 2007 involved 100 cloned embryos, implanted into 50 female macaque monkeys, none of which gave rise to a viable pregnancy. In January 2008, scientists at Stemagen, a stem cell research and development company in California, announced that they had cloned five human embryos by means of SCNT and that the embryos had matured to the stage at which they could have been implanted in a womb. However, the scientists destroyed the embryos after five

A SHEEP NAMED DOLLY

Dolly was a female Finn Dorset sheep that lived from 1996 to 2003, the first successfully cloned mammal, produced by Scottish geneticist Ian Wilmut and colleagues of the Roslin Institute, near Edinburgh. The announcement in February 1997 of the world's first clone of an adult animal was a milestone in science, dispelling decades of presumption that adult mammals could not be cloned and igniting a debate concerning the many possible uses and misuses of mammalian cloning technology.

Clones had been generated previously in the laboratory, but only from embryonic cells or from the adult cells of plants and "lower" animals such as frogs. Decades of attempts to clone mammals from existing adults had met with repeated failure, which led to the presumption that something special and irreversible must happen to the DNA of mammalian cells during the animal's development. Indeed, until 1997 it had been generally accepted dogma that adult mammalian cells are no longer genetically totipotent, or capable of giving rise to all of the different cell and tissue types (e.g., liver, brain, and bone) required for making a complete and viable animal. That Dolly remained alive and well long after her birth—that she had a functional heart, liver, brain, and other organs, all derived genetically from the nuclear DNA of an adult mammary-gland cell—proved otherwise. At the very minimum, the specific tissue from which Dolly's nuclear DNA was derived must have been totipotent. By extension, it was reasonable to suggest that the nuclear DNA of other adult tissues also remains totipotent. With the successful creation of Dolly, this speculation became a testable hypothesis.

Dolly did not spring from the laboratory bench fully formed but developed to term normally in the uterus of a

Scottish Blackface ewe. Although Dolly's nuclear genome was derived from a mammary-gland cell taken from an adult Finn Dorset ewe, that nucleus had to be fused by electrical pulses with an unfertilized egg cell, the nucleus of which had been removed. The "host" egg cytoplasm was taken from a Scottish Blackface ewe, and later another Scottish Blackface ewe served as the surrogate mother. Furthermore, in order for the mammary gland cell nucleus and genomic DNA to be accepted and functional within the context of the host egg, the donor cell first had to be induced to abandon the normal cycle of growth and division and enter a quiescent stage. To do this, researchers deliberately withheld nutrients from the cells. The importance of this step had been determined experimentally, and although a number of hypotheses had been raised to explain its necessity, which, if any, of them was correct remained unclear. Nevertheless, starting with a collection of donor cell nuclei and host egg cytoplasms, a number of fused couplets successfully formed embryos; these were transferred to surrogate ewes. Of 13 recipient ewes, one became pregnant, and 148 days later, which is essentially normal gestation for a sheep, Dolly was born.

On Feb. 14, 2003, Dolly was euthanized by veterinarians after being found to suffer from progressive lung disease. Her body was preserved and displayed at the National Museum of Scotland in Edinburgh.

days, presumably because of ethical reasons and the need to perform molecular analyses on the embryos.

Therapeutic Cloning

Therapeutic cloning is intended to use cloned embryos for the purpose of extracting stem cells from them, without ever implanting the embryos in a womb. Therapeutic cloning enables the cultivation of stem cells that are genetically identical to a patient. The stem cells could be stimulated to differentiate into any of the more than 200 cell types in the human body. The differentiated cells then could be transplanted into the patient to replace diseased or damaged cells without the risk of rejection by the immune system. These cells could be used to treat a variety of conditions, including Alzheimer's disease, Parkinson's disease, diabetes mellitus, stroke, and spinal cord injury. In addition, stem cells could be used for in vitro (laboratory) studies of normal and abnormal embryo development or for testing drugs to see if they are toxic or cause birth defects.

Although stem cells have been derived from the cloned embryos of animals such as mice, the generation of stem cells from cloned primate embryos has proved exceptionally difficult. For example, in 2007 stem cells successfully derived from cloned macaque embryos were able to differentiate into mature heart cells and brain neurons. However, the experiment started with 304 egg cells and resulted in the development of only two lines of stem cells, one of which had an abnormal Y chromosome. Likewise, the production of stem cells from human embryos has been fraught with the challenge of maintaining embryo viability. In 2001 scientists at Advanced Cell Technology, a research company in Massachusetts, successfully transferred DNA from human cumulus cells, which are cells that cling to and nourish human eggs, into eight enucleated eggs. Of these eight eggs, three developed into early-stage embryos (containing four to six cells); however, the embryos survived only long enough to divide once or twice. In 2004 South Korean researcher Hwang Woo Suk claimed to have cloned human embryos using SCNT and to have extracted stem cells from the embryos. However, this later proved to be a fraud; Hwang had fabricated evidence and had actually carried out the process of parthenogenesis, in which an unfertilized egg begins to divide with only half a genome. The following year, a team of researchers from the University of Newcastle upon Tyne was able to grow a cloned human embryo to the 100-cell blastocyst stage using DNA from embryonic stem cells. Although they did not generate a line of stem cells from the blastocyst, this was a major step forward in therapeutic cloning research. However, because embryonic stem cells can only be isolated and cloned from a recently fertilized embryo and not a grown patient, the procedure was considered to have only narrow clinical application.

Progress in research on therapeutic cloning in humans has been slow relative to the advances made in reproductive cloning in animals. This is primarily due to the technical challenges and ethical controversy arising from the procuring of human eggs solely for research purposes and the lack of understanding of how somatic cell DNA is reprogrammed into an undifferentiated embryonic state during SCNT. In addition, the development of induced pluripotent stem cells, which are derived from somatic cells that have been reprogrammed to an embryonic state through the introduction of specific genetic factors into the cell nuclei, has challenged the use of cloning methods and of human eggs.

Chapter 6: Military Technology

Inventions have also contributed to the grimmest of human endeavours, the deliberate and organized application of deadly force against other people. Indeed, so powerful has been the thirst for victory, or the fear of defeat, that no effort has been spared to devise ever more powerful implements of war. In this string of efforts, we have moved far beyond the use of handheld weapons and defenses to the splitting and fusing of atoms to create explosions of almost unimaginable power.

THE SPEAR

Though early man probably employed spears of fire-hardened wood, spearheads of knapped stone were used long before the emergence of any distinction between hunting and military weapons. Bronze spearheads closely followed the development of alloys hard enough to keep a cutting edge and represented, with the piercing ax, the earliest significant military application of bronze. Spearheads were also among the earliest militarily significant applications of iron, no doubt because existing patterns could be directly extrapolated from bronze to iron. Though the hafting is quite different, bronze Sumerian spearheads of the 3rd millennium BCE differ only marginally in shape from the leaf-shaped spearheads of classical Greece.

The spears of antiquity were relatively short, commonly less than the height of the warrior, and typically were wielded with one hand. As defensive armour and other weapons of shock combat (notably the sword) improved, spear shafts were made longer and the use of the spear became more specialized. The Greek hoplite's spear was about 9 feet (2.7 m) long; the Macedonian *sarissa* was twice that length in the period of Alexander's conquests and it grew to some 21 feet (6.3 m) in Hellenistic times.

Javelins, or throwing spears, were shorter and lighter than spears designed for shock combat and had smaller heads. The distinction between javelin and spear was slow to develop, but by classical times the heavy spear was clearly distinguished from the javelin, and specialized javelin troops were commonly used for skirmishing. A throwing string was sometimes looped around the shaft and tied to the thrower's finger to impart spin to the javelin on release. This improved the weapon's accuracy and probably increased the range and penetrating power by permitting a harder cast.

A significant refinement of the javelin was the Roman pilum. The pilum was relatively short, about 5 feet (1.5 m) long, and had a heavy head of soft iron that made up nearly one-third of the weapon's total length. The weight of this weapon restricted its range but gave it greater impact. Its head of soft iron was intended to bend on impact, preventing an enemy from throwing it back.

Like the spear, the javelin was relatively unaffected by the appearance of iron and retained its characteristic form until it was finally abandoned as a serious weapon in the 16th century.

THE BOW AND ARROW

The bow was simple in concept, yet it represented an extremely sophisticated technology. In its most basic form, the bow consisted of a stave of wood slightly bent by the tension of a bowstring connecting its two ends. The bow stored the force of the archer's draw as potential energy, then transferred it to the bowstring as kinetic energy, imparting velocity and killing power to the arrow. The bow could store no more energy than the archer was capable of producing in a single movement of the muscles of his back and arms, but it released the stored energy at a higher velocity, thus overcoming the arm's inherent limitations.

Assuming the same length of draw and available force, the total amount of potential energy that an archer could store in a bow was a function of the bow's length; that is, the longer the arms of the bow, the more energy stored per unit of work expended in the draw and, therefore, the more kinetic energy imparted to the string and arrow. The disadvantage of a long bow was that the stored energy had to serve not only to drive the string and arrow but also to accelerate the mass of the bow itself. Because the longer bow's more massive arms accelerated more slowly, a longer bow imparted kinetic energy to the string and arrow at a lower velocity. A shorter bow, on the other hand, stored less energy for the same amount of work expended in the draw, but it compensated for this through its ability to transmit the energy to the arrow at a higher velocity. In sum, the shorter bow imparted less total energy to the arrow, but it did so at a higher velocity. Therefore, in practice maximum range was attained by a short, stiff bow shooting a very light arrow, and maximum killing power at medium ranges was attained by a long bow driving a relatively heavy arrow.

The Early Bow

The simple bow, made from a single piece of wood, was known to Neolithic hunters; it is clearly depicted in cave paintings of 30,000 BCE and earlier. The first improvement was the reflex bow, a bow that was curved forward,

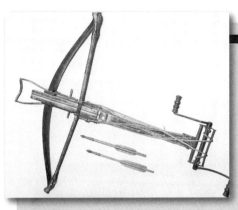

Stirrup crossbow, French, 14th century. Courtesy of the West Point Museum Collections, U.S. Military Academy

THE CROSSBOW

The crossbow was an important technical achievement of the Middle Ages that enjoyed the great distinction of being outlawed (at least for use against Christians) by the Lateran Council of 1139. Its origins are obscure, but its earliest appearance in Europe was in the technologically advanced Italian cities of the 10th and 11th centuries. The destructive power that made it so feared derived from the substitution of metal (wrought iron or mild steel) for wood in its construction. The metal bow, relatively short, was bent by one of two methods. In the earlier version the bowman placed the weapon on the ground, bow down and stock, or cross, upright. Setting his foot in a stirrup in the lower part of the stock, he bent over, caught the bowstring in a hook suspended from his belt, straightened up, and brought the string into the locking device in the groove of the stock. In the second version the stirrup was still used, but a hand crank or winch drew back the string. A small lever triggered the release of the short bolt, or quarrel, which was capable of piercing chain mail and had a range of up to 1,000 feet (300 m).

Despite the introduction of the English (or Welsh) longbow, with its quicker rate of fire, the crossbow continued its reign as the supreme hand missile weapon until, and even for a long time after, the introduction of firearms. The crossbow's great advantage was that no particular strength

> was needed to wield it effectively. In addition to its power, it owed its long success to its versatility (it could be fired from a reclining position or from behind a parapet) and its less bulky ammunition. The slower rate of fire (compared with the longbow) may also have been an advantage in some situations. Not until the late 15th century did it definitely give way to the harquebus, a portable but very heavy matchlock gun.

or reflexively, near its centre so that the string lay close against the grip before the bow was drawn. This increased the effective length of the draw since it began farther forward, close to the archer's left hand.

The next major improvement, one that was to remain preeminent among missile weapons until well into the modern era, was the composite recurved bow. This development overcame the inherent limitations of wood in stiffness and tensile strength. The composite bow's resistance to bending was increased by reinforcing the rear, or belly, of the bow with horn; its speed and power in recoil were increased by overlaying the front of the bow with sinew, usually applied under tension. The wooden structure of this composite thus consisted of little more than thin wooden strips supporting the horn and sinew. The more powerful composite bows, being very highly stressed, reversed their curvature when unstrung. They acquired the name "recurved" since the outer arms of the bow curved away from the archer when the bow was strung, which imparted a mechanical advantage at the end of the draw. Monumental and artistic evidence suggest that the principle of the composite recurved bow was known as early as 3000 BCE.

A prime advantage of the composite bow was that it could be engineered to essentially any desired strength. By following the elaborate but empirically understood trade-off between length and stiffness referred to above, the bowyer (a maker of bows) could produce a short bow capable of propelling light arrows to long ranges, a long, heavy bow designed to maximize penetrative power at relatively short ranges, or any desired compromise between.

Arrows

Arrow design was probably the first area of military technology in which production considerations assumed overriding importance. As a semi-expendable munition that was used in quantity, arrows could not be evaluated solely by their technological effectiveness; production costs had to be considered as well. As a consequence, the materials used for arrowheads tended to be a step behind those used for other offensive technologies. Arrowheads of flint and obsidian, knapped to remarkably uniform standards, survived well into the Bronze Age, and bronze arrowheads were used long after the adoption of iron for virtually every other military cutting or piercing implement.

Arrow shafts were made of relatively inexpensive wood and reed throughout history, though considerable labour was involved in shaping them. Remarkably refined techniques for fastening arrowheads of flint and obsidian to shafts were well in hand long before recorded history. (The importance of arrow manufacturing techniques is reflected in the survival in modern English of the given name Fletcher, the title of a specialist in attaching feathers to the arrow shaft.)

GUNPOWDER

Few inventions have had an impact on human affairs as dramatic and decisive as that of gunpowder. The development of a means of harnessing the energy released by a chemical reaction in order to drive a projectile against a target marked a watershed in the harnessing of energy to human needs. Before gunpowder, weapons were designed around the limits of their users' muscular strength; after gunpowder, they were designed more in response to tactical demand.

Chinese alchemists discovered the recipe for what became known as black powder in the 9th century CE; this was a mixture of finely ground potassium nitrate (also called saltpetre), charcoal, and sulfur in approximate proportions of 75:15:10 by weight. The resultant gray powder behaved differently from anything previously known; it exploded on contact with open flame or a red-hot wire, producing a bright flash, a loud report, dense white smoke, and a sulfurous smell. It also produced considerable quantities of superheated gas, which, if confined in a partially enclosed container, could drive a projectile out of the open end. The Chinese used the substance in rockets, in pyrotechnic projectors much like Roman candles, in crude cannon, and, according to some sources, in bombs thrown by mechanical artillery. This transpired long before gunpowder was known in the West, but development in China stagnated. The development of black powder as a tactically significant weapon was left to the Europeans, who probably acquired it from the Mongols in the 13th century (though diffusion through the Arab Muslim world is also a possibility).

Nineteenth-century experiments revealed sharp differences in the amount of gas produced by charcoal burned from different kinds of wood. For example, dogwood charcoal decomposed with potassium nitrate was found to yield nearly 25 percent more gas per unit weight than fir, chestnut, or hazel charcoal and some 17 percent more than willow charcoal. These scientific observations confirmed the insistence of early—and thoroughly unscientific—texts that charcoal from different kinds of wood was suited to different applications. Willow charcoal, for example, was preferred for cannon powder and dogwood charcoal for small arms—a preference substantiated by 19th-century tests. For all this, the empirically derived recipe for gunpowder was fixed during the 14th century and hardly varied thereafter. Subsequent improvements were almost entirely concerned with the manufacturing process and with the ability to purify and control the quality of the ingredients.

Between 1870 and 1890 much work was done on the development of propellants and explosives. Nitrocellulose-based powders (called ballistite in France and cordite in Britain) became the standard propellant. Whereas black powder produced a large quantity of solid material upon combustion, quickly fouling barrels and pouring out huge clouds of smoke, nitrocellulose produced mostly gas and was therefore labeled "smokeless powder." Also, it produced three times the energy of black powder and burned at a more controllable rate. Such characteristics made possible a shift to longer and smaller-diameter projectiles. Bore diameters of shoulder weapons were reduced to calibres of about .30 inch, or 7.5 to 8 mm. Muzzle velocities ranged from 2,000 to 2,800 feet (600 to 850 m) per second, and accurate range extended to 1,000 yards (914 m) and beyond. Because lead projectiles were too soft to be used at such increased power and velocity, they were sheathed in harder metal. In 1881 a Swiss officer, Eduard Alexander Rubin, was the first to perfect a full-length, copper-jacketed bullet.

Most forms of gunpowder produced today are either single-base (i.e., consisting of nitrocellulose alone) or double-base (consisting of a combination of nitrocellulose and nitroglycerin). Both types are prepared by plasticizing nitrocellulose with suitable solvents, rolling it into thin sheets, and cutting the sheets into small squares called granules or grains, which are then dried. Control of the burning rate is achieved by varying the composition, size, and geometric shape of the propellant grains, and sometimes by surface treatment or coating of the grains. Generally the goal is to produce a propellant that is slowly converted to gas in the initial stages of burning and more rapidly converted as burning progresses.

Rifled Muzzle-Loaders

During the 17th and 18th centuries, smoothbore muzzle-loading infantry muskets became the dominant force in war. Nevertheless, as killing machines they were relatively inefficient. Their heavy, round lead balls delivered bone-crushing and tissue-destroying blows when they hit a human body, but beyond 75 yards (68 m) even trained infantrymen found it difficult to hit an individual adversary. Volley fire against massed troops delivered effective projectiles out to 200 yards (182 m), but at 300 yards (274 m) balls from muzzle-loaders lost most of their lethality. Also, while well-trained soldiers could load and shoot their muskets five times per minute, volley fire led to a collective rate of only two to three shots per minute.

Early Rifling

These ballistic shortcomings were a product of the requirement that the projectile, in order to be quickly rammed from muzzle to breech, had to fit loosely in the barrel.

When discharged, it wobbled down the barrel, contributing to erratic flight after it left the muzzle. Rifled barrels, in which spiral grooves were cut into the bore, were known to improve accuracy by imparting a gyroscopic spin to the projectile, but reloading rifled weapons was slowed because the lead ball had to be driven into the barrel's rifling. Greased cloth or leather patches eased the problem somewhat, but the rate of fire of rifles was still much lower than that of smoothbore muskets.

One possible solution was the creation of mechanisms that allowed the bullet to be loaded at the breech instead of the muzzle. Many such ideas were tested during the 18th century, but, given the craftsman-based manufacture of the day, none was suited to large-scale production. Special army units in Europe and America used rifled muzzle-loaders, such as the flintlock British Baker rifle, to harass the enemy at long ranges, while most infantrymen continued to carry muzzle-loading smoothbores. For this reason, inventors concentrated on adapting rifled barrels to muzzle-loaders. In 1826 Henri-Gustave Delvigne of France, seeking a means of expanding the projectile without making it difficult to ram home, created a narrow powder chamber at the breech end of the barrel against which a loosely fitting lead ball came to rest. Ramrod blows expanded the soft lead at the mouth of the chamber so that, when fired, the bullet fit the rifling tightly. In 1844 another French officer, Louis-Étienne de Thouvenin, introduced yet a better method for expanding bullets. His *carabine à tige* embodied a post or pillar (*tige*) at the breech against which the bullet was expanded.

Minié Rifles

These rifles worked better than earlier types, but their deformed balls flew with reduced accuracy. Captain

Claude-Étienne Minié, inspired by Delvigne's later work with cylindrical bullets, designed longer, smaller-diameter projectiles. These had the same weight as larger round balls, but possessed greater cross-sectional density and therefore retained their velocity better. Moreover, while the flat base of Minié's projectile was deformed against the pillar as in Thouvenin's weapon, the rest of the bullet maintained its shape and accuracy. The French army combined these ideas in the Carabine Modèle 1846 à tige and the Fusil d'Infanterie Modèle 1848 à tige.

In order to combat the tendency of muzzle-loading rifles to become difficult to load as gunpowder residue collected in the barrels, Minié suggested a major simplification — eliminating the pillar and employing in its place a hollow-based bullet with an iron expander plug that caused the projectile to engage the rifling when the weapon was fired. This new projectile could be loaded into dirty rifles with ease, and, because it was not deformed while loading, it had greater accuracy.

Officials in several countries, notably Britain and the United States, saw the significance of Minié's invention. In 1851 the Royal Small Arms Factory, Enfield, embarked upon production of the .702-inch (18-mm) Pattern 1851 Minié rifle. In the Crimean War (1854–56), Russian troops armed with smoothbore muskets were no match for Britons shooting P/51 rifles. Massed formations were easy prey, as were cavalry and artillery units. A correspondent for the *Times of London* wrote: "The Minié is king of weapons... the volleys of the Minié cleft [Russian soldiers] like the hand of the Destroying Angel."

Swiss experiments demonstrated that an expander plug was not necessary when a bullet's side walls were thin enough, and the British designed a smaller-calibre rifle using this type of Minié bullet. The result was a .577-

inch (15-mm) weapon firing "cylindro-conoidal" projectiles—essentially a lead cylinder with a conical nose. "Enfield" as a weapon name was first generally applied to these Pattern 1853 rifles. Subsequent tests indicated that rifles with 33-inch (84-cm) barrels could provide accuracy equal to the 39-inch (99-cm) P/53 barrels. When the resulting P/53 Short Rifles were issued, there began a century-long trend toward shorter weapons.

In the United States, experiments undertaken in the late 1840s led to the adoption of a .58-inch Minié-type bullet and a family of arms designed to use it. The Model 1855 rifled musket, with a 40-inch (102-cm) barrel, produced a muzzle velocity of 950 feet (290 m) per second. All Model 1855 weapons used mechanically operated tape priming, intended to eliminate the manual placement of percussion caps on the nipple, but this system proved too fragile and was eliminated with the introduction of a simplified Model 1861 rifled musket. During the American Civil War (1861–65), the Union government purchased both Model 1861 and Model 1863 rifled muskets as its basic infantry weapon. In the Confederacy, domestic production was supplemented by purchases of Enfield P/53 rifles and other European weapons.

The Civil War clearly demonstrated the deadly effect of rifled muskets, although many battlefield commanders only slowly appreciated the changing nature of warfare. Individual soldiers could hit their opposing numbers with accurate fire out to 250 yards, so that frontal assaults, in which soldiers advanced in neat ranks across open fields, had to be abandoned. By 1862 both sides were building field entrenchments and barricades to provide protection from rifle and artillery fire.

THE SUBMARINE

Dutch inventor Cornelis Drebbel is usually credited with building the first submarine, an oar-powered "diving boat" that he demonstrated in the River Thames about 1620. A number of submarine boats were conceived over the following two centuries, ranging from the *Turtle*, a one-man craft invented by David Bushnell during the American Revolution, to the *H.L. Hunley*, a modified iron boiler powered by eight men cranking a propeller that attacked a Union ship during the American Civil War and then sank with all hands aboard.

A major limitation of all the early submarines was their lack of a suitable means of propulsion. In the second half of the 19th century, several attempts were made using steam power and compressed air, but practical propulsion was not possible until the development of the electric motor. The submarine *Nautilus*, built in 1886, was an all-electric craft. Propelled by two 50-horsepower electric motors operated from a 100-cell storage battery, it achieved a surface speed of six knots (7 miles, or 11 km, per hour). But the battery had to be recharged and overhauled at short intervals, and the craft was never able to travel more than 80 miles (129 km) without a battery recharge. In France, Gustave Zédé launched the *Gymnote* in 1888; it, too, was propelled by an electric motor and was extremely maneuverable but tended to go out of control when it dived.

The end of the 19th century was a period of intensive submarine development, and Zédé collaborated in a number of designs sponsored by the French navy. A most successful French undersea craft of the period was the *Narval*, designed by Maxime Laubeuf, a marine engineer in the navy. Launched in 1899, the *Narval* was a double-hulled craft, 111.5 feet (34 m) long, propelled on the

MILITARY TECHNOLOGY

DAVID BUSHNELL'S *TURTLE*

The first submarine to be put to military use was built in 1775 for use against British warships during the American Revolution. A pear-shaped vessel known to posterity as the *Turtle*, this one-man submersible vessel was made of oak reinforced with iron bands and measured about 7.5 feet (2.3 m) long by 6 feet (1.8 m) wide. It was designed to be propelled under water by an operator who turned its propeller by hand. The craft was armed with a mine, or

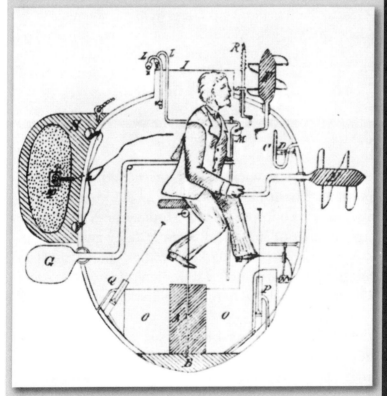

Bushnell's submarine torpedo boat, 1776. Drawing of a cutaway view made by Lieutenant Commander F. M. Barber in 1885 from a description left by Bushnell. Courtesy of the U.S. Navy

"torpedo," to be attached to the hull of an enemy ship. In 1776, in New York harbour, the *Turtle* tried to sink the British warship HMS *Eagle* but failed; none of its succeeding missions were successful.

The *Turtle*'s inventor, David Bushnell, was born in 1742 in Saybrook, Conn. Graduated from Yale in 1775, at the outbreak of the American Revolution, Bushnell went back to Saybrook, where he built his unique vessel. Though the submarine gave proof of underwater capability, the attacks were failures, partly because Bushnell's physical frailty made it almost impossible for him to perform in person the many demanding functions required to control the craft. Gen. George Washington, however, gave him a commission in the engineers, where he rose to captain and command of the U.S. Army Corps of Engineers stationed at West Point. In his later years Bushnell studied medicine and entered practice in Warrenton, Ga., where he died in 1824.

surface by a steam engine and by electric motors when submerged. The ballast tanks were located between the double hulls, a concept still in use today. The *Narval* made a large number of successful dives. Further French progress in submarines was marked by the four Sirène-class steam-driven undersea craft completed in 1900–01 and the *Aigrette*, completed in 1905, the first diesel-driven submarine of any navy.

Similarly, there were submarine successes in the United States by rival inventors John P. Holland (an Irish immigrant) and Simon Lake. Holland launched his first under-sea craft in 1875. This one and its successors were significant in combining water ballast with horizontal rudders for diving. In 1895, in competition with Swedish designer Torsten Nordenfelt, Holland received an order from the U.S. Navy for a submarine. This was to be the *Plunger*, propelled by steam on the surface and by electricity when submerged. The craft underwent many design changes and finally was abandoned before completion. Holland returned the funds advanced by the navy and built his next submarine (his sixth) at his own expense. This was the *Holland*, a 53.25-foot (16.25-m) craft launched in 1897 and accepted by the navy in 1900. For underwater propulsion the *Holland* had an electric motor, and it was propelled on the surface by a gasoline engine. The submarine's armament consisted of a bow torpedo tube, for which three torpedoes were carried, and two dynamite guns. With its nine-man crew the *Holland* was a successful boat; it was modified many times to test different arrangements of propellers, diving planes, rudders, and other equipment.

Holland's chief competitor, Simon Lake, built his first submarine, the *Argonaut I*, in 1894; it was powered by a gasoline engine and electric motor. This and Lake's other early boats were intended as undersea research craft. In 1898 the *Argonaut I* sailed from Norfolk, Va., to New York City under its own power, predating the cruises of the French *Narval* and marking the first time an undersea craft operated extensively in the open sea. Lake's second submarine was the *Protector*, launched in 1901.

Of the major naval powers at the turn of the century, only Britain remained indifferent toward submarines. Finally, in 1901, the Royal Navy ordered five of the *Holland*-design undersea craft. Germany completed its first submarine, the *U-1* (for *Unterseeboot 1*), in 1905. This craft was 139 feet (42.25 m) long, powered on the surface by a heavy oil engine and by an electric motor when submerged, and was armed with one torpedo tube. Thus, the stage was set for the 20th-century submarine, a craft propelled on the surface by diesel engines and underwater by battery-powered electric motors, submerging by diving planes and taking on water ballast, and armed with torpedoes for sinking enemy ships.

THE MACHINE GUN

Self-actuated machine guns, which operated under energy generated by a fired round, became militarily effective at the end of the 19th century, after the introduction of nitrocellulose-based gunpowder. Smokeless powder burned at a more controlled rate than did the older black-powder propellants, generating pressures that built up over a longer time. The first automatic weapons to take advantage of this were heavy guns firing the high-velocity metal-jacketed cartridges designed for the new bolt-action infantry rifles.

RECOIL

The first successful automatic machine gun was invented by Hiram Stevens Maxim, an American working in Europe. Beginning about 1884, he produced a number of weapons in which the bullet's recoil energy was employed to unlock the breechblock from the barrel, to extract and eject the fired case from the gun, and to store sufficient energy in a main spring to push the bolt forward, pick up a fresh round, load the chamber, and lock the piece. Both barrel and breechblock, locked together, recoiled a short distance to the rear; then the barrel was stopped and the block continued back alone. If the trigger was held in firing position, the weapon would continue to fire until it expended all of its ammunition. Rounds were fed to the gun on belts, which could be clipped together to provide continuous fire, and overheating was solved by surrounding the barrel in a metal jacket in which water was circulated from a separate container.

German infantrymen operating a Maxim machine gun during World War I.
Imperial War Museum

Maxim's salesmen provided armies with guns in any calibre, usually matching their current rifle cartridge. In Britain, Maxim guns were first chambered for the .45-inch Martini-Henry cartridge, but, as issued in 1891, they fired the .303-inch smokeless-powder round of the Lee-Metford rifle. During the Russo-Japanese War (1904–05), the Russians used English-made Maxim guns chambered for their 7.62-mm Mosin-Nagant round. Their Model 1910 weighed about 160 pounds (72 kg), including mount, water-cooling apparatus, and a protective steel shield for the gunner. The German Model 1908, chambered for the 7.92-mm Mauser cartridge, weighed 100 pounds (45 kg) with its sled mount. Such light weights, made possible because the cartridge was the sole source of power, allowed these weapons to be operated by special infantry units.

Machine guns of the Maxim type had a destructive power never seen before in warfare. In the 1890s, British infantry units used Maxim guns, fabricated under contract by Vickers Sons, to cut down hordes of poorly armed rebels in Africa and Afghanistan. In World War I, a few of them could cause thousands of casualties. Their defensive fire so limited the offensive power of infantry that the entire Western Front, from the Swiss border to the English Channel, became one vast siege operation.

GAS OPERATION

Not all the early heavy machine guns were of the recoil-operated Maxim type. Gas operation was also employed. In this system a piston located in a cylinder below the barrel was driven to the rear by gas diverted from the barrel through a port. The piston unlocked the breechblock and sent the bolt back against the main spring. A new round was then picked up, moved into the chamber, and fired on the forward stroke.

The best-known gas-operated heavy machine gun was the Hotchkiss, introduced in France in 1892 and modified several times until the definitive version of 1914. It was air-cooled, but the barrel itself was heavy and provided with metal fins to increase heat radiation. A slower method of feeding ammunition by short strips instead of long belts also helped to keep the weapon from overheating. The Japanese used Hotchkiss guns chambered for their 6.5-mm round against Russia in 1904–05. In World War I, two French Hotchkiss guns firing 8-mm Lebel cartridges were said to have fired 75,000 rounds each in the defense of Verdun and to have remained serviceable.

BLOWBACK

A third principle of machine-gun operation was often called "blowback." In this, the action and barrel were never locked rigidly together; the barrel did not move, nor was there a gas cylinder and piston. To prevent the breech from opening so early that propellant gases would rupture the spent cartridge case, the block was heavy and the main spring strong. Also, there was usually a linkage of parts not quite on centre to delay the actual opening. Finally, the barrel was shorter than usual, allowing the bullet and gases to leave the barrel quickly.

The Austrian Schwarzlose of 1907/12, firing eight-mm Mannlicher rounds, operated by delayed blowback. It was entirely satisfactory in combat during World War I.

THE ASSAULT RIFLE

At the beginning of the 20th century, the ballistic performance of infantry rifles was tailored to the long-range requirements of a bygone era when foot soldiers demanded weapons that could reach great distances to halt the dreaded cavalry charge. Beginning early in World War I, however, battlefields became no-man's-lands pockmarked by shell craters and crisscrossed by miles of barbed-wire entanglements, and machine guns dominated the 1,000 or more yards (914 or more metres) between trench lines. Rifles were shot at those extreme ranges, but they could not equal the destructive power of artillery and machine guns. They were also too cumbersome and powerful for offensive assaults on enemy trenches. The greater mobility of troops accompanying armoured vehicles reinforced the need for lighter, more portable weapons of improved effectiveness at close quarters.

Such changing conditions of war led to experiments with automatic weapons firing rounds of lower velocity or lighter weight. One result, which saw its first use in World War I, was a new weapon called the machine carbine or submachine gun. Derived from the semiautomatic pistol and firing pistol-calibre ammunition with muzzle velocities of only about 1,000 feet (300 m) per second, submachine guns were fitted with shoulder stocks (and sometimes forward hand grips). Such weapons offered easier handling than rifles while providing greater accuracy and more rapid fire than most handguns.

Under the pressures of World War II, the major powers used millions of submachine guns, but after the war, the submachine gun lost importance as a military weapon. With an effective range limited to about 200 yards (182 m), it could not fill the broad gap between the low-power pistol cartridge and the full-power rifle cartridge. This gap, which constituted the ground upon which modern infantrymen found themselves fighting, had to be filled by another new weapon, which would fire a cartridge of intermediate power.

A hint at this new weapon had been given during World War I, when Vladimir Grigorevich Fyodorov, father of Russian automatic weapons, married the 6.5-mm cartridge of the Japanese Arisaka rifle to an automatic rifle. In 1916 he unveiled his new weapon, the Avtomat Fyodorova. Owing to the turmoil of the Russian Revolution of 1917, only about 3,200 of Fyodorov's weapons were delivered. Nevertheless, they pointed the way to future infantry weapon design.

During World War II, Hugo Schmeisser designed a light rifle to fire the Germans' 7.92-mm Kurz, or "Short," cartridge, which was of the same calibre as the Mauser rifle cartridge but was lighter and shorter and was therefore of a less potent, "intermediate" power. The weapon, known variously as the MP43, MP44, or Sturmgewehr ("Assault Rifle") 44, was loaded by a curved box magazine holding 30 rounds and was designed for most effective fire at about 300 yards (274 m). Only 425,000 to 440,000 of these rifles were built—too few and too late for the German war effort—but they were based on a concept that would dominate infantry weapons for the rest of the century.

Late in the war the Soviets also began a search for a rifle to shoot their 7.62-mm intermediate cartridge, which produced a muzzle velocity of 2,330 feet (710 m) per second. Historical evidence suggests that they were influenced by the Sturmgewehr, but to what extent remains uncertain. In 1947 they adopted a weapon designed by Mikhail Timofeyevich Kalashnikov, naming it the Avtomat Kalashnikova. Like the German weapon, the AK-47 was operated by diverting some of the propellant gases into a cylinder above the barrel; this drove a piston that forced the bolt back against its spring and cocked the hammer for the next round. At the turn of a selector switch, the action could be changed from semiautomatic to fully automatic, firing at a rate of 600 rounds per minute. The AK-47 was made of forged and milled steel, giving it a weight of 10.6 pounds (4.8 kg) with a loaded 30-round magazine. The receiver of the AKM version, introduced in 1959, was made of lighter sheet metal, reducing the weight to 8.3 pounds, and the AK-74 version, following later trends in the West, switched to a 5.45-mm cartridge.

Kalashnikov's assault rifles became the most significant infantry weapons of the post-World War II era. In many variants, they were adopted and made by countries all over the world. Between 30 and 50 million AKs were produced within four decades of the series' introduction, more than any other firearm in history.

After the Korean War (1950–53), U.S. military researchers dissatisfied with rifle-power ammunition began to test a .22-inch (5.56-mm) cartridge that propelled a light projectile at a high muzzle velocity of 3,000 feet (900 m) per second. To fire this "small-calibre, high-velocity" round, in 1958 they chose the AR-15 rifle, designed by Eugene M.

Stoner for the ArmaLite Division of Fairchild Engine and Airplane Corporation. The AR-15 was gas-operated, but it eliminated the piston in favour of a tube that directed propellant gases directly into an expansion chamber between the bolt and bolt carrier. By reducing the number of working parts and chambering the rifle for a smaller cartridge, Stoner had come up with a lightweight weapon that, even on automatic fire, produced a manageable recoil and yet was capable of inflicting fatal wounds at 300 yards (274 m) and beyond. In 1961 the air force purchased the AR-15, renaming it the M16. Six years later, with units in Vietnam finding the weapon very effective under the close conditions of jungle warfare, the army adopted it as the M16A1.

After U.S. troops in Europe were issued the M16, a series of trials ensued that ended with the decision, in 1980, to adopt a standard 5.56-mm NATO cartridge. West Germany introduced the G41, a 5.56-mm version of the G3, and Belgium replaced the FAL with the FNC. The British and French armies developed new assault rifles with compact "bullpup" designs, in which the bolt, receiver, and magazine were behind the handgrip and trigger and much of the shoulder stock was occupied by the operating mechanism. This permitted a much shorter weapon than orthodox designs, in which the magazine and receiver were ahead of the trigger. As a result, the French FA MAS and British L85A1 were only some 30 to 31 inches (76 to 78.5 cm) long—compared with the M16, which was 39 inches (99 cm) overall. Many of the newer models were built with lightweight plastic shoulder stocks and magazines, as well as receivers made of aluminum.

THE TANK

Tanks are essentially tracked, protected weapon platforms that make the weapons mounted in them more effective by their cross-country mobility and by the protection they provide for their crews. The use of vehicles for fighting dates to the 2nd millennium BCE, when horse-drawn war chariots were used in the Middle East by the Egyptians, Hittites, and others as mobile platforms for combat with bows and arrows. The concept of protected vehicles can be traced back through the wheeled siege towers and battering rams of the Middle Ages to similar devices used by the Assyrians in the 9th century BCE. The two ideas began to merge in the battle cars proposed in 1335 by Guido da Vigevano, in 1484 by Leonardo da Vinci, and by others, down to James Cowen, who took out a patent in England in 1855 for an armed, wheeled, armoured vehicle based on the steam tractor.

But it was only at the beginning of the 20th century that armoured fighting vehicles began to take practical form. By then the basis for them had become available with the appearance of the traction engine and the automobile. Thus, the first self-propelled armoured vehicle was built in 1900 in England when John Fowler & Company armoured one of their steam traction engines for hauling supplies in the South African (Boer) War (1899–1902). The first motor vehicle used as a weapon carrier was a powered quadricycle on which F. R. Simms mounted a machine gun in 1899 in England. The inevitable next step was a vehicle that was both armed and armoured. Such a vehicle was constructed to the order of Vickers, Son and Maxim, Ltd. and was exhibited in London in 1902. Two years later a fully armoured car with a turret was built in France by the Société Charron, Girardot et Voigt, and another was built concurrently in Austria by the Austro-Daimler Company.

To complete the evolution of the basic elements of the modern armoured fighting vehicle, it remained only to adopt tracks as an alternative to wheels. This became inevitable with the appearance of the tracked agricultural tractor, but there was no incentive for this until after the outbreak of World War I in 1914, which radically changed the situation. Its opening stage of mobile warfare accelerated the development of armoured cars, numbers of which were quickly improvised in Belgium, France, and Britain. The ensuing trench warfare, which ended the usefulness of armoured cars, brought forth new proposals for tracked armoured vehicles. Most of these resulted from attempts to make armoured cars capable of moving off roads, over broken ground, and through barbed wire. The first tracked armoured vehicle was improvised in July 1915, in Britain, by mounting an armoured car body on a Killen-Strait tractor. The vehicle was constructed by the Armoured Car Division of the Royal Naval Air Service, whose ideas, backed by the First Lord of the Admiralty, Winston S. Churchill, resulted in the formation of an Admiralty Landships Committee. A series of experiments by this committee led in September 1915 to the construction of the first tank, called "Little Willie." A second model, called "Big Willie," quickly followed. Designed to cross wide trenches, it was accepted by the British Army, which ordered 100 tanks of this type (called Mark I) in February 1916.

Simultaneously but independently, tanks were also developed in France. Like the very first British tank, the first French tank (the Schneider) amounted to an armoured box on a tractor chassis; 400 were ordered in February 1916. But French tanks were not used until April 1917, whereas British tanks were first sent into action on Sept. 15, 1916. Only 49 were available, and their success was limited, but on Nov. 20, 1917, 474 British tanks were concentrated at the Battle of Cambrai and achieved a spectacular breakthrough. These tanks, however, were too slow and had too short an operating range to exploit the

British Mark I tank with antibomb roof and "tail," 1916. Courtesy of the Imperial War Museum, London; photographs, Camera Press

breakthrough. In consequence, demand grew for a lighter, faster type of tank, and in 1918 the 14-ton Medium A appeared with a speed of 8 miles (13 km) per hour and a range of 80 miles (128 km). After 1918, however, the most widely used tank was the French Renault F.T., a light 6-ton vehicle designed for close infantry support.

When World War I ended in 1918, France had produced 3,870 tanks, and Britain 2,636. Most French tanks survived into the postwar period; these were the Renault F.T., much more serviceable than their heavier British counterparts. Moreover, the Renault F.T. fitted well with traditional ideas about the primacy of the infantry, and the French army adopted the doctrine that tanks were a mere auxiliary to infantry. France's lead was followed in most other countries; the United States and Italy both assigned tanks to infantry support and copied the Renault

F.T. The U.S. copy was the M1917 light tank, and the Italian the Fiat 3000. The only other country to produce tanks by the end of the war was Germany, which built about 20.

THE BALLISTIC MISSILE

Ballistic missiles are rocket-propelled weapons that travel by momentum in a high, arcing trajectory after they have been launched into flight by a brief burst of power. Land-based ballistic missiles can be divided into two general categories according to their range: intermediate-range ballistic missiles (IRBMs) and intercontinental ballistic missiles (ICBMs). IRBMs have ranges of about 600 to 3,500 miles (960 to 5,600 km), while ICBMs have ranges exceeding 3,500 miles. Most modern submarine-launched ballistic missiles (SLBMs) are of intermediate range.

THE V-2

The precursor of modern ballistic missiles was the German V-2, a single-stage, fin-stabilized missile propelled by liquid oxygen and ethyl alcohol to a maximum range of about 200 miles (321 km). The V-2 was officially designated the A-4, being derived from the fourth of the *Aggregat* series of experiments conducted at Kummersdorf and Peenemunde under General Walter Dornberger and the civilian scientist Wernher von Braun.

The most difficult technical problem facing the V-2 was achieving maximum range. An inclined launch ramp was normally used to give missiles maximum range, but this could not be used with the V-2 because the missile was quite heavy at liftoff (more than 12 tons) and would not be traveling fast enough to sustain anything approaching horizontal flight. Also, as the rocket used up its fuel its weight (and velocity) would change, and this had to be allowed for in the aiming. For these reasons the V-2 had to be launched straight up and then had to change to the flight angle that would give it maximum range. The Germans calculated this angle to be slightly less than 50°.

The change in direction mandated some sort of pitch control during flight, and, because a change in pitch would induce yaw, control was needed on the yaw axis, too. Added to these problems was the natural tendency of a cylinder to rotate. Thus, the V-2 (and every ballistic missile afterward) needed a guidance and control system to deal with in-flight rolling, pitching, and yawing. Using three-axis autopilots adapted from German aircraft, the V-2 was controlled by large vertical fins and smaller stabilizing surfaces to dampen roll and by vanes attached to the horizontal fins to modify pitch and yaw. Vanes were also installed in the exhaust nozzle for thrust vector control.

A combination of in-flight weight changes and changes in atmospheric conditions presented additional problems. Even over the fairly limited course of a V-2 trajectory (with a range of approximately 200 miles [320 km] and an altitude of roughly 50 miles [80 km]), changes in missile velocity and air density produced drastic shifts in the distance between the centre of gravity and the centre of aerodynamic pressure. This meant the guidance system had to adjust its input to the control surfaces as the flight proceeded. As a result, V-2 accuracy never ceased to be a problem for the Germans.

Still, the missile caused a great deal of damage. The first V-2 used in combat was fired against Paris on Sept. 6, 1944. Two days later the first of more than 1,000 missiles was fired against London. By the end of the war 4,000 of these missiles had been launched from mobile bases against Allied targets. During February and March 1945, only weeks before the war in Europe ended, an average of 60 missiles was launched weekly. The V-2 killed an estimated five people per launch, and several V-2 attacks killed more than 100 people. There was no known defense against the V-2; it could not be intercepted and, traveling faster than sound, it arrived unexpectedly. The V-2 threat was eliminated only by bombing the launch sites and forcing the German army to retreat beyond missile range.

The V-2 obviously ushered in a new age of military technology. After the war there was intense competition between the United States and the Soviet Union to obtain these new missiles, as well as to obtain the German scientists who had developed them. The United States succeeded in capturing both Dornberger and von Braun as well as more than 60 V-2s; it was not revealed precisely what (or whom) the Soviets captured. However, given the relative immaturity of ballistic missile technology at that time, neither country achieved usable ballistic missiles for some time. During the late 1940s and early 1950s most of the nuclear competition between the two countries dealt with strategic bombers. Events in 1957, however, reshaped this contest.

The First ICBMs

In 1957 the Soviets launched a multistage ballistic missile (later given the NATO designation SS-6 Sapwood) as well as the first man-made satellite, Sputnik. This prompted the "missile gap" debate in the United States and resulted in higher priorities for the U.S. Thor and Jupiter IRBMs. Although originally scheduled for deployment in the early 1960s, these programs were accelerated, with Thor being deployed to England and Jupiter to Italy and Turkey in 1958. Thor and Jupiter were both single-stage, liquid-fueled missiles with inertial guidance systems and warheads of 1.5 megatons. Political difficulties in deploying these missiles on foreign soil prompted the United States to develop

MILITARY TECHNOLOGY

Titan II rocket, lifting off from an underground silo. Developed as an intercontinental ballistic missile, the Titan II also served as a launch vehicle for the Gemini manned spacecraft missions and military and civilian satellites. U.S. Air Force; photograph provided by Donald Boelling

THE TITAN I AND TITAN II

Titan was a series of U.S. rockets that were originally developed as ICBMs but subsequently became important expendable space-launch vehicles. Titan I, the first in the series, was built by Martin Company (later Lockheed Martin Corporation) for the U.S. Air Force in the late 1950s. A two-stage ICBM fueled by kerosene and liquid oxygen, it was designed to deliver a 4-megaton nuclear warhead to targets in the Soviet Union more than 5,000 miles (8,000 km) away. Between 1962 and 1965 several squadrons of Titan Is were operational at air force bases in the western United States. The missiles were stored underground in reinforced-concrete silos but had to be raised to ground level for launch and required a minimum of 15 to 20 minutes for fueling.

By 1965 Titan I had been replaced by Titan II, a much larger ICBM (approximately 100 feet [30 m] long) that could be launched directly from its silo and was fueled by internally stored hypergolic fuels (self-igniting liquids such as hydrazine and nitrogen tetroxide). Tipped with a nine-megaton warhead—the most powerful nuclear explosive ever mounted on a U.S. delivery vehicle—and stationed at bases in the central and western United States, Titan II was the principal weapon in the land-based U.S. nuclear arsenal

> until it was replaced by more-accurate solid-fueled ICBMs such as Minuteman. The last Titan IIs were deactivated between 1982 and 1987. Converted Titan IIs were used by the National Aeronautics and Space Administration (NASA) as launchers for Gemini manned spacecraft during the 1960s. After its deactivation as an ICBM, Titan II was modified by Lockheed Martin to launch satellites for U.S. government use.

ICBMs, so that by late 1963 Thor and Jupiter had been terminated. (The missiles themselves were used extensively in the space program.)

The Soviet SS-6 system was an apparent failure. Given its limited range (less than 3,500 miles [5,600 km]), it had to be launched from northern latitudes in order to reach the United States. The severe weather conditions at these launch facilities (Novaya Zemlya and the Arctic mainland bases of Norilsk and Vorkuta) seriously degraded operational effectiveness; pumps for liquid propellants froze, metal fatigue was extreme, and lubrication of moving parts was nearly impossible. In 1960 a missile engine exploded during a test, killing Mitrofan Ivanovich Nedelin, chief of the Strategic Rocket Forces, and several hundred observers.

Possibly as a result of these technical failures (and possibly in response to the deployment of Thor and Jupiter), the Soviets attempted to base the SS-4 Sandal—an IRBM with a 1-megaton warhead and a range of 900–1,000 miles (1,440–1,600 km)—closer to the United States and in a warmer climate. This precipitated the Cuban missile crisis of 1962, after which the SS-4 was withdrawn to Central Asia. (It was unclear whether the United States' deactivation of Thor and Jupiter was a condition of this withdrawal.)

In the meantime, the United States was developing operational ICBMs to be based on U.S. territory. The first versions were the Atlas and the Titan I. The Atlas-D (the first version deployed) had a liquid-fueled engine that generated 360,000 pounds (1,601 kilonewtons) of thrust. The missile was radio-inertial guided, launched above ground, and had a range of 7,500 miles (12,000 km). The follow-on Atlas-E/F increased thrust to 390,000 pounds (1,734 kilonewtons), used all-inertial guidance, and moved from an aboveground to horizontal canister launch in the E and, finally, to silo-stored vertical launch in the F. The Atlas E carried a 2-megaton, and the Atlas F a 4-megaton, warhead. The Titan I was a two-stage, liquid-fueled, radio-inertial guided, silo-launched ICBM carrying a 4-megaton warhead and capable of traveling 6,300 miles (10,000 km). Both systems became operational in 1959.

NUCLEAR WEAPONS

Nuclear weapons are devices that produce enormous explosive energy as a result of nuclear fission, nuclear fusion, or a combination of the two processes. Fission weapons are commonly referred to as atomic bombs. Fusion weapons are referred to as thermonuclear bombs or, more commonly, hydrogen bombs. The first atomic bombs were built by the United States during World War II and were dropped on Hiroshima and Nagasaki, Japan, in 1945. After the war American scientists developed the first hydrogen bomb, which was tested in 1952.

The First Atomic Bombs

In early January 1939, Austrian physicist Otto Frisch rushed to Copenhagen to inform the Danish scientist

Niels Bohr of an important discovery: German scientists had caused a uranium nucleus to fission, or break apart into two smaller pieces, by bombarding the uranium with low-speed neutrons. Bohr was about to leave for a visit to the United States, where he reported the news to colleagues. The revelation set off experiments at many laboratories, and nearly 100 articles were published about the exciting phenomenon by the end of the year. Bohr postulated that the uranium isotope uranium-235 was the one undergoing fission; the other isotope, uranium-238, merely absorbed the neutrons. In addition, it was discovered that neutrons were also produced during the fission process; on average, each fissioning atom produced more than two neutrons. If the proper amount of material were assembled, these free neutrons might create a chain reaction. Under special conditions, a very fast chain reaction might produce a very large release of energy—in short, a weapon of fantastic power might be feasible.

Producing a Controlled Chain Reaction

The possibility that an atomic bomb might first be developed by Nazi Germany alarmed many scientists and was drawn to the attention of U.S. Pres. Franklin D. Roosevelt by Albert Einstein, then living in the United States. The president appointed an Advisory Committee on Uranium, which reported on Nov. 1, 1939, that a chain reaction in uranium was possible, though unproved. In March 1940 it was confirmed that the isotope uranium-235 was indeed responsible for low-speed neutron fission in uranium. The Advisory Committee on Uranium increased its support of chain-reaction experiments and arranged for a study of possible methods for separating the uranium-235 isotope from the much more abundant uranium-238. Two separation processes were explored: the centrifuge process, in which the heavier isotope is spun to the outside, and the gaseous diffusion process, in which gaseous uranium hexafluoride is diffused through barriers, or filters.

During the summer of 1940, researchers at the University of California at Berkeley discovered a new element, element 93. They named the new element neptunium and inferred that this element would decay into yet another new element, element 94, which in turn might also fission under neutron bombardment. Element 94 was discovered on Feb. 23, 1941; during the following year the researchers named it plutonium and established that it did indeed undergo fission—and at a rate much higher than that of uranium-235. Plutonium became recognized as another possible fuel for an atomic bomb, in addition to uranium-235.

In May 1941 a review committee reported that a nuclear explosive probably could not be available before 1945. A chain reaction in natural uranium was probably 18 months off, and it would take at least an additional year to produce enough plutonium and three to five years to separate enough uranium-235 for a bomb. In the fall of 1941 the chain-reaction experiments at Columbia University in New York City yielded negative results. All subsequent work was transferred to the University of Chicago, where the world's first controlled nuclear chain reaction was achieved by Italian physicist Enrico Fermi and his group on Dec. 2, 1942, in the squash court under the stands of the university's football field.

The Manhattan Project

The United States' entry into World War II in December 1941 was decisive in providing funds for a massive research and production effort for obtaining fissionable materials. The U.S. Army Corps of Engineers was given the job of constructing the production plants, and in August 1942 the project was officially given the name Manhattan

The first atomic bomb test, near Alamogordo, N.M., July 16, 1945. Jack Abbey/Los Alamos National Laboratory

Engineer District—hence Manhattan Project, the name by which this effort would be known ever afterward. Gen. Leslie R. Groves, selected to head the project, chose the three key production sites—Oak Ridge, Tenn.; Los Alamos, N.M.; and Hanford, Wash.—and selected the large corporations to build and operate the atomic factories. In December contracts were signed with the DuPont Company to design, construct, and operate the plutonium production reactors and to develop the plutonium separation facilities. Two types of factories to enrich uranium were built at Oak Ridge.

In November 1942 Groves and physicist J. Robert Oppenheimer visited the Los Alamos Ranch School, some 60 miles (100 km) north of Albuquerque, N.M., and Groves approved it as the site for the main scientific laboratory, often referred to by its code name Project Y. Oppenheimer would be the scientific director of the laboratory, where the design, development, and final manufacture of the weapon would take place.

Racing to Build the Bombs

The emphasis during the summer and fall of 1943 was on the gun method of assembly, in which a projectile, a subcritical piece of uranium-235 (or plutonium-239), would be placed in a gun barrel and fired into a target, another subcritical piece. After the mass was joined (and now supercritical), a neutron source would be used to start the chain reaction. A problem developed with applying the gun method to plutonium, however. Some neutrons would always be present in a plutonium assembly and would cause it to explode prematurely, producing comparatively little energy. The problem was overcome by the invention of a second gun design—the implosion method, in which a solid sphere of plutonium would be compressed and brought to a critical state by imploding "lenses" of high explosive surrounding the sphere.

By 1944 the Manhattan Project was spending money at a rate of more than $1 billion per year. Delays meant almost certainly that the war in Europe would be over before the weapon could be ready. The ultimate target changed from Germany to Japan. After Roosevelt's death on April 12, 1945, Pres. Harry S. Truman was given an extensive briefing on the status of the project: the uranium-235 gun design had been finalized, but a sufficient quantity of uranium-235 would not be accumulated until about August 1. Enough plutonium-239 would be available for an implosion assembly to be tested in early July; a second would be ready in August. Several dozen B-29 bombers had been modified to carry the weapons, and construction of a staging base was under way at Tinian, in the Mariana Islands, 1,500 miles (2,400 km) south of Japan.

The test of the plutonium weapon was named Trinity; it was fired at 5:29:45 AM on July 16, 1945, at the Alamogordo Bombing Range in south-central New Mexico. The theorists' predictions of the energy release, or yield, of the device ranged from the equivalent of less than 1,000 tons of TNT to the equivalent of 45,000 tons (that is, from 1 to 45 kilotons of TNT). The test actually produced a yield of about 21,000 tons.

Col. Paul W. Tibbets Jr., pilot of the Enola Gay, *the plane that dropped the atomic bomb on Hiroshima, Japan, Aug. 6, 1945.* U.S. Air Force photograph

The Weapons are Used

A single B-29 bomber named *Enola Gay* flew over Hiroshima, Japan, on Monday, Aug. 6, 1945, at 8:15 AM. The untested uranium-235 gun-assembly bomb, nicknamed Little Boy, was airburst 1,900 feet (580 m) above the city to maximize destruction; it was later estimated to yield 15 kilotons. Two-thirds of the city area was destroyed. The population present at the time was estimated at 350,000; of these, 140,000 died by the end of the year. The second weapon, a duplicate of the plutonium-239 implosion assembly tested in Trinity and nicknamed Fat Man, was

On Aug. 9, 1945, three days after detonating a uranium-fueled atomic bomb over Hiroshima, Japan, the United States dropped a plutonium-fueled atomic bomb over the Japanese port of Nagasaki. U.S. Department of Defense

dropped on August 9 over Nagasaki, where at 11:02 AM the weapon was airburst at 1,650 feet (500 m); it was later estimated that the explosion yielded 21 kilotons. About half of Nagasaki was destroyed, and about 70,000 of some 270,000 people present at the time of the blast died by the end of the year.

THE FIRST HYDROGEN BOMBS

U.S. research on thermonuclear weapons was started by a conversation in September 1941 between Fermi and Hungarian-American physicist Edward Teller. Fermi wondered if the explosion of a fission weapon could ignite a mass of deuterium sufficiently to begin nuclear fusion. (Deuterium is an isotope of hydrogen with one proton and one neutron in the nucleus.) Teller presented his findings to a group of theoretical physicists, who suggested that the use of tritium be investigated as well. (Tritium is an isotope of hydrogen with one proton and two neutrons in the nucleus.) When the Los Alamos laboratory was being planned, a small research program on the Super, as the thermonuclear design came to be known, was included, though the more urgent task of developing a fission weapon always took precedence.

THE "SUPER BOMB" IS APPROVED

In the fall of 1945, after the success of the atomic bomb and the end of World War II, the future of the Manhattan Project, including Los Alamos and the other facilities, was unclear. In April 1946 a conference led by Teller at Los Alamos reviewed the status of the Super. One central design problem was how to ignite the thermonuclear fuel. It was recognized early on that a mixture of deuterium and tritium theoretically could be ignited at lower temperatures and would have a faster reaction time than deuterium alone, but the question of how to achieve ignition remained unresolved. To resolve it would be a formidable task, and the limited resources of Los Alamos could not support an extensive program.

The first Soviet test of an atomic bomb in 1949 stimulated an intense secret debate about whether to proceed with the hydrogen bomb project. One of the strongest statements against proceeding with the "super bomb" came from a committee chaired by Oppenheimer, of which a majority believed "that the extreme dangers to mankind inherent in the proposal wholly outweigh any military advantages." Nevertheless, the Joint Chiefs of Staff, State Department, Defense Department, Joint Committee on Atomic Energy, and a special subcommittee of the National Security Council all recommended proceeding. In January 1950, Truman directed that work continue on all forms of nuclear weapons, including hydrogen bombs.

THE WEAPON IS BUILT

In the months that followed Truman's decision, researchers at Los Alamos struggled with the problem of building a thermonuclear weapon, particularly the problem of how to ignite the fuel. After months of discouraging calculations, breakthroughs were achieved between February and April 1951. One breakthrough was the recognition that the burning of thermonuclear fuel would be more efficient if a high density were achieved throughout the fuel prior to raising its temperature. A second breakthrough was the recognition that these conditions—high compression and high temperature throughout the fuel—could be achieved by containing and converting the radiation from an exploding fission weapon and then using this energy to compress a separate component containing the thermonuclear fuel. The major figures in these breakthroughs were Teller and Polish-American mathematician

The first thermonuclear weapon (hydrogen bomb), code-named Mike, was detonated at Enewetak atoll in the Marshall Islands, Nov. 1, 1952. Three of a series of photographs taken at an altitude of 12,000 feet (3,600 m) 50 miles (80 km) from the detonation site. U.S. Air Force photograph

Stanislaw M. Ulam. The two-stage radiation implosion design proposed in their reports, which led to the modern concept of thermonuclear weapons, became known as the Teller-Ulam configuration.

Without hesitation, Teller and Ulam's colleagues adopted the new design, and in September 1951 Los Alamos proposed a test of the Teller-Ulam concept, to be code-named Mike. Teller and Ulam's theoretical ideas were transformed into a device that weighed 82 tons, in part because of cryogenic (low-temperature) refrigeration equipment necessary to keep the deuterium in liquid form. It was successfully detonated on Nov. 1, 1952, at Enewetak atoll in the western Pacific. The explosion achieved a yield of 10.4 megatons (million tons), 500 times larger than the Nagasaki bomb, and it produced a crater 6,240 feet (1,900 m) in diameter and 164 feet (50 m) deep.

Chapter 7: Observation and Measurement

Magnitude—especially the magnitude of time and distance—is a property that has always defied human beings but that people in turn have always tried to master. And so we have the invention of clocks and calendars for dividing time into discrete units, or of telescopes and microscopes for viewing objects too small or too far away for the eye to see. The inventions profiled here, unlike those profiled elsewhere in this book, do not attempt to move, manipulate, or change our surrounding world; rather, they attempt to help us quantify and comprehend it.

THE GREGORIAN CALENDAR

The calendar now in general worldwide use is known as the Gregorian calendar, having been proclaimed in 1582 by Pope Gregory XIII as a reform of the Julian calendar, which had been in use for some 1,600 years, since the reign of Julius Caesar. The new calendar had its origin in the desire for a solar calendar that kept in step with the seasons and possessed fixed rules of intercalation, that is, the insertion of days (such as leap-year days) into a calendar to bring it in line with the year of the seasons (tropical year). Because it was developed in Western Christendom, the new calendar had also to provide a method for dating movable religious feasts such as Easter, the timing of which had been based on a lunar reckoning.

The Julian calendar year of 365.25 days was too long, since the correct value for the tropical year is 365.242199 days. This error of 11 minutes 14 seconds per year amounted to almost one and a half days in two centuries, and seven days in 1,000 years, and as a result the calendar became increasingly out of phase with the seasons. From time to time, the problem was placed before church councils, but no action was taken because the astronomers who were consulted doubted whether enough precise information was available for a really accurate value of the tropical year to be obtained.

By 1545, however, the vernal equinox, which was used in determining Easter, had moved an intolerable 10 days from its proper date; and in December, when the Council of Trent met for the first of its sessions, it authorized Pope Paul III to take action to correct the error. Correction required a solution, however, that neither Paul III nor his successors were able to obtain in satisfactory form until nearly 1572, the year of election of Pope Gregory XIII. Gregory found various proposals awaiting him and agreed to issue a papal bull, or decree, that the Jesuit astronomer Christopher Clavius (1537–1612) began to draw up, using suggestions made by the astronomer and physician Luigi Lilio (also known as Aloysius Lilius; died 1576).

The papal bull *Inter gravissimas* ("In the gravest concern") was issued on Feb. 24, 1582. First, in order to bring the vernal equinox back to March 21, the day following the Feast of St. Francis (that is, October 5) was to become October 15, thus omitting 10 days. Second, to bring the year closer to the true tropical year, a value of 365.2422 days was accepted. This value differed by 0.0078 days per year from the Julian calendar reckoning, amounting to 0.78 days per century, or 3.12 days every 400 years. It was therefore promulgated that three out of every four centennial years should be common years, that is, not leap years; and this practice led to the rule that no centennial years should be leap years unless exactly divisible by 400.

Thus, 1700, 1800, and 1900 were not leap years, as they would have been in the Julian calendar, but the year 2000 was.

The Gregorian calendar firmly established January 1 as the beginning of its year, and it was widely referred to as the New Style calendar, with the Julian becoming the Old Style. The reform was well received by such astronomers as Johannes Kepler and Tycho Brahe and by the Catholic princes of Europe. However, because of the division between the Eastern and Western Christian churches and between Protestants and Roman Catholics, the obvious advantages of the Gregorian calendar were not accepted everywhere, and in some places adoption was extremely slow. In France, Italy, Luxembourg, Portugal, and Spain, the New Style calendar was adopted in 1582, and it was in use by most of the German Roman Catholic states as well as by Belgium and part of the Netherlands by 1584. Switzerland's change was gradual, on the other hand, beginning in 1583 and being completed only in 1812. Hungary adopted the New Style in 1587, and then there was a pause of more than a century before the first Protestant countries made the transition from the Old Style calendar. In 1699–1700, Denmark and the Dutch and German Protestant states embraced the New Style, although the Germans declined to adopt the rules laid down for determining Easter. The Germans preferred to rely instead on astronomical tables and specified the use of the *Tabulae Rudolphinae* (1627; "Rudolphine Tables"), based on the 16th-century observations of Tycho Brahe. They acceded to the Gregorian calendar rules for Easter only in 1776. Sweden adopted the New Style in 1753, although the Swedes, because they had in 1740 followed the German Protestants in using their astronomical methods for determining Easter, declined to adopt the Gregorian calendar rules until 1844. Japan adopted the New Style in

1873; Egypt adopted it in 1875; and between 1912 and 1917 it was accepted by Albania, Bulgaria, China, Estonia, Latvia, Lithuania, Romania, and Turkey. The now-defunct Soviet Union adopted the New Style in 1918, and Greece in 1923.

In Britain and the British dominions, the change was made in 1752, when the difference between the New and Old Style calendars amounted to 11 days: the lag was covered by naming the day after Sept. 2, 1752, as Sept. 14, 1752. There was widespread misunderstanding among the public, however, even though legislation authorizing the change had been framed to avoid injustice and financial hardship. The Alaskan territory retained the Old Style calendar until 1867, when it was transferred from Russia to the United States.

THE CLOCK

A clock is a machine for displaying time in which a device that performs regular movements in equal intervals is linked to a counting mechanism that records the number of movements. All clocks, of whatever form, are made on this principle.

The origin of the all-mechanical escapement clock is unknown; the first such devices may have been invented and used in monasteries to toll a bell that called the monks to prayers. The first mechanical clocks to which clear references exist were large, weight-driven machines fitted into towers and known today as turret clocks. These early devices struck only the hours and did not have hands or a dial.

The oldest surviving clock in England is that at Salisbury Cathedral, which dates from 1386. A clock erected at Rouen, France, in 1389 is still extant, and one built for Wells Cathedral in England is preserved in the

Science Museum in London. The Salisbury clock strikes the hours, and those of Rouen and Wells also have mechanisms for chiming at the quarter hour. These clocks are large, iron-framed structures driven by falling weights attached to a cord wrapped around a drum and regulated by a mechanism known as a verge (or crown wheel) escapement. Their errors probably were as large as a half hour per day. The first domestic clocks were smaller wall-mounted versions of these large public clocks. They appeared late in the 14th century, and few examples have survived; most of

Reconstruction of the waterpowered mechanical clock built under the direction of Su Sung, 1088 CE. By John Christiansen after Joseph Needham, et al. Courtesy of the East Asian History of Science Library, Cambridge, and Cambridge University Press

them, extremely austere in design, had no cases or means of protection from dust.

About 1450, clock makers working probably in southern Germany or northern Italy began to make small clocks driven by a spring. These were the first portable timepieces, representing an important landmark in horology. The time-telling dials of these clocks usually had an hour hand only (minute hands did not generally appear until the 1650s) and were exposed to the air; there was normally no form of cover such as a glass until the 17th century, though the mechanism was enclosed, and the cases were made of brass.

About 1581 the great Italian scientist Galileo noticed the characteristic timekeeping property of the pendulum. The Dutch astronomer and physicist Christiaan Huygens was responsible for the practical application of the pendulum as a time controller in clocks from 1656 onward. Huygens's invention brought about a great increase in the importance and extent of clock making. Clocks, weight-driven and with short pendulums, were encased in wood and made to hang on the wall, but these new eight-day wall clocks had very heavy weights, and many fell off weak plaster walls and were destroyed. The next step was to extend the case to the floor, and the grandfather clock was born. In 1670 the long, or seconds, pendulum was introduced by English clock makers with the anchor escapement.

THE WATCH

The first portable timepieces appeared shortly after 1500, early examples being made by Peter Henlein, a locksmith in Nürnberg, Ger. The escapement used in the early watches was the same as that used in the early clocks, the verge.

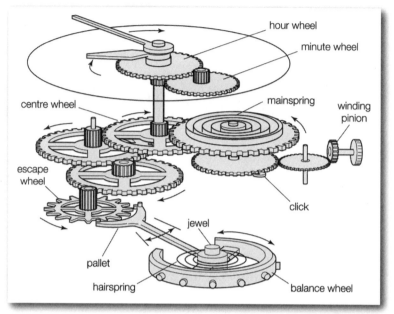

Typical components of a mechanical watch. Encyclopædia Britannica, Inc.

Early watches were made notably in Germany and at Blois in France, among other countries, and were generally carried in the hand or worn on a chain around the neck. They usually had only one hand for the hours.

One of the main defects of the early watches was the variation in the torque exerted by the mainspring. The mainspring is the element that drives the watch; it consists of a flat spring-steel band stressed in bending or coiling. When the watch, or other spring-driven mechanism, is wound, the curvature of the spring is increased, and energy is thus stored. This energy is transmitted to the oscillating section of the watch (called the balance) by the wheeltrain and escapement, the motion of the balance itself controlling the release of the escapement and consequently the timing of the watch. In early watches, the force of the mainspring was greater when fully wound than when it was almost run down. Since the timekeeping of a watch fitted with a verge escapement was greatly influenced by the force driving it, this problem was quite serious. Solution of the problem was advanced almost as soon as the mainspring was invented (about 1450) by the application of the fusee, a cone-shaped, grooved pulley used together with a barrel containing the mainspring. With this arrangement, the mainspring was made to rotate a barrel in which it was housed; a length of catgut, later replaced by a chain, was wound on it, the other end being coiled around the fusee. When the mainspring was fully wound, the gut or chain pulled on the smallest radius of the cone-shaped fusee; as the mainspring ran down, the leverage was progressively increased as the gut or chain pulled on a larger radius. With correct proportioning of mainspring and fusee radii, an almost constant torque was maintained as the mainspring unwound.

Controlling the oscillations of a balance with a spring was an important step in the history of timekeeping. English physicist Robert Hooke designed a watch with a balance spring in the late 1650s; there appears to be no evidence, however, that the spring was in the form of a spiral, a crucial element that would become widely employed. Dutch scientist Christiaan Huygens was probably the first to design (1674–75) a watch with a spiral balance spring. The balance spring is a delicate ribbon of steel or other suitable spring material, generally wound into a spiral form. The inner end is pinned into a collet (a small collar), which fits friction-tight on the balance staff, while the outer end is held in a stud fixed to the movement. This spring acts on the balance as gravity does on the pendulum. If the balance is displaced to one side, the spring is wound and energy stored in it; this energy is then restored to the balance, causing it to swing nearly the same distance to the other side if the balance is released.

If there were no frictional losses (e.g., air friction, internal friction in the spring material, and friction at the pivots), the balance would swing precisely the same distance to the other side and continue to oscillate indefinitely; because of these losses, however, the oscillations in practice die away. It is the energy stored in the mainspring and fed to the balance through the wheel train and escapement that maintains the oscillations.

THE TELESCOPE

Galileo is credited with having developed telescopes for astronomical observation in 1609. While the largest of his instruments was only about 48 inches (120 cm) long and had an objective diameter of 2 inches (5 cm), it was equipped with an eyepiece that provided an upright (i.e., erect) image. Galileo used his modest instrument to explore such celestial phenomena as the valleys and mountains of the Moon, the phases of Venus, and the four largest Jovian satellites, which had never been systematically observed before.

Galileo's instrument was a refracting telescope, one in which the incoming light is bent, or refracted toward a focal point on passing through the lens. The reflecting telescope is an instrument in which a mirror reflects the light back to a focal point instead of refracting it. This type of telescope was developed in 1668 by the great English mathematician and physicist Sir Isaac Newton, though the Scottish astronomer James Gregory had independently conceived of an alternative reflector design in 1663. The Frenchman Laurent Cassegrain introduced another variation of the reflector in 1672. Near the end of the century, others attempted to construct refractors as

THE HALE TELESCOPE

The George Ellery Hale Telescope, one of the world's largest and most powerful reflecting telescopes, is located at the Palomar Observatory, Mount Palomar, Calif. Financed by the Rockefeller Foundation, the telescope at Palomar was completed in 1948 and named in honour of the noted American astronomer George Ellery Hale, who supervised the designing of the instrument.

The main mirror of the Hale Telescope measures 200 inches (508 cm) across and weighs 14.5 tons. It is made of Pyrex, carefully ground and polished to the correct curvature and coated with aluminum to give a durable, highly reflective surface. The entire movable part of the telescope weighs more than 500 tons, yet it is so smoothly supported and delicately balanced on hydrostatic bearings that a $1/12$-horsepower motor can turn it to follow the apparent rotation of the sky. The tube, of open-girder construction, is 60 feet (18 m) long. The telescope was the first in the world to have an observing cage for the astronomer inside the telescope tube at prime focus; in the 1960s observations made from this cage determined that quasars were the most distant objects in the universe.

long as 200 feet (61 m), but these instruments were too awkward to be effective.

The most significant contribution to the development of the telescope in the 18th century was that of Sir William Herschel. Herschel, whose interest in telescopes was kindled by Gregory's modest 2-inch (5-cm) reflector design, persuaded the king of England to finance the construction of a reflector with a 40-foot (12-m) focal length and a 47-inch (120-cm) mirror. Herschel is credited with having used this instrument to lay the observational groundwork for the concept of extragalactic "nebulas"— i.e., galaxies outside the Milky Way system.

Reflectors continued to evolve during the 19th century with the work of William Parsons, 3rd Earl of Rosse, and William Lassell. In 1845 Lord Rosse constructed in Ireland a reflector with a 73-inch (185-cm) mirror and a focal length of about 52 feet (16 m). For 75 years this telescope ranked as the largest in the world and was used to explore thousands of nebulas and star clusters. Lassell built several reflectors, the largest of which was on Malta; this instrument had a 49-inch (124-cm) primary mirror and a focal length of more than 33 feet (10 m). His telescope had greater reflecting power than that of Parsons, and it enabled him to catalog 600 new nebulas as well as to discover several satellites of the outer planets—Triton (Neptune's largest moon), Hyperion (Saturn's 8th moon), and Ariel and Umbriel (two of Uranus's moons).

Refractor telescopes, too, underwent development during the 18th and 19th centuries. The last significant one to be built was the 40-inch (1-m) refractor at Yerkes Observatory. Installed in 1897, it remains the largest refracting system in the world. Its objective was designed and constructed by the optician Alvan Clark, while the mount was built by the firm of Warner & Swasey.

The reflecting telescope predominated in the 20th century. The rapid proliferation of larger and larger instruments of this type began with the installation of the 100-inch (254-cm) reflector at the Mount Wilson Observatory near Pasadena, Calif. The technology for mirrors underwent a major advance when the Corning Glass Works (in Steuben County, N.Y.) developed Pyrex. This borosilicate glass, which undergoes substantially less expansion than ordinary glass, was used in the 200-inch

(508-cm) Hale reflector built in 1950 at the Palomar Observatory. Pyrex also was utilized in the main mirror of the world's largest telescope, the 236-inch (6-m) reflector of the Special Astrophysical Observatory in Zelenchukskaya, Russia.

THE MICROSCOPE

The concept of magnification has long been known. About 1267 English philosopher Roger Bacon wrote in *Perspectiva*, "[We] may number the smallest particles of dust and sand by reason of the greatness of the angle under which we may see them," and in 1538 Italian physician Girolamo Fracastoro wrote in *Homocentrica*, "If anyone should look through two spectacle glasses, one being superimposed on the other, he will see everything much larger."

Three Dutch spectacle makers—Hans Jansen, his son Zacharias Jansen, and Hans Lippershey—have received credit for inventing the compound microscope about 1590. The first portrayal of a microscope was drawn about 1631 in the Netherlands. It was clearly of a compound microscope, with an eyepiece and an objective lens. This kind of instrument, which came to be made of wood and cardboard, often adorned with polished fish skin, became increasingly popular in the mid-17th century and was used by the English natural philosopher Robert Hooke to provide regular demonstrations for the new Royal Society. These demonstrations commenced in 1663, and two years later Hooke published a folio volume titled *Micrographia*, which introduced a wide range of microscopic views of familiar objects (fleas, lice, and nettles among them). In this book he coined the term *cell*.

Hidden in the unnumbered pages of *Micrographia*'s preface is a description of how a single high-powered lens

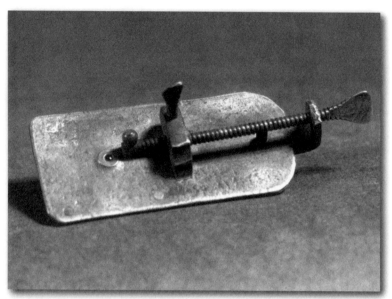

Microscope made by Antonie van Leeuwenhoek. © Comstock Images/ Jupiterimages

could be made into a serviceable microscope, and it was using this design that the Dutch civil servant Antonie van Leeuwenhoek began his pioneering observations of freshwater microorganisms in the 1670s. He made his postage-stamp-sized microscopes by hand, and the best of them could resolve details around 0.7 m. His fine specimens discovered in excellent condition at the Royal Society more than three centuries later prove what a great technician he was. Using his simple microscope, Leeuwenhoek effectively launched microbiology in 1674, and single-lensed microscopes remained popular until the 1850s. In 1827 they were used by Scottish botanist Robert Brown to demonstrate the ubiquity of the cell nucleus, a term he coined in 1831.

Simple microscopes using single lenses can generate fine images; however, they can also produce spurious colours due to chromatic aberration, in which different wavelengths of light do not come to the same focus. The aberrations were worse in the compound microscopes of the time, because the lenses magnified the aberrations at least as much as they magnified the images. Although the compound microscopes were beautiful objects that conferred status on their owners, they produced inferior images. In 1733 the amateur English optician Chester Moor Hall found by trial and error that a combination of a convex crown-glass lens and a concave flint-glass lens could help to correct chromatic aberration in a telescope, and in 1774 Benjamin Martin of London produced a pioneering set of colour-corrected lenses for a microscope.

The appearance of new varieties of optical glasses encouraged continued development of the microscope in the 19th century, and considerable improvements were made in understanding the geometric optics of image formation. The concept of an achromatic (non-colour-distorting) microscope objective was finally introduced in 1791 by Dutch optician Francois Beeldsnijder, and the English scientist Joseph Jackson Lister in 1830 published a work describing a theoretical approach to the complete design of microscope objectives. The physics of lens construction was examined by German physicist Ernst Abbe. In 1868 he invented an apochromatic system of lenses, which had even better colour correction than achromatic lenses, and in 1873 he published a comprehensive analysis of lens theory. Light microscopes were produced in the closing quarter of the 19th century that reached the effective limits of optical microscopy. Subsequent instruments—such as phase-contrast microscopes, interference microscopes, and confocal microscopes—solved specific problems that had previously made studying important objects such as living cells very difficult.

RADAR

Serious developmental work on radar ("radio detecting and ranging") began in the 1930s, but the basic idea of radar had its origins in the classical experiments on electromagnetic radiation conducted by German physicist Heinrich Hertz during the late 1880s. Hertz set out to verify experimentally the earlier theoretical work of Scottish physicist James Clerk Maxwell, which had led to the conclusion that radio waves can be reflected from metallic objects and refracted by a dielectric medium, just as light waves can. Hertz demonstrated these properties in 1888, using radio waves at a wavelength of 66 cm (which corresponds to a frequency of about 455 megahertz [MHz]).

The potential utility of Hertz's work as the basis for the detection of targets of practical interest did not go unnoticed at the time, but there was no economic, societal, or military need for radar until the early 1930s, when long-range military bombers capable of carrying large payloads were developed. This prompted the major countries of the world to look for a means with which to detect the approach of hostile aircraft. The United States, Great Britain, Germany, France, the Soviet Union, Italy, the Netherlands, and Japan all began experimenting with radar within about two years of one another and embarked, with varying degrees of motivation and success, on its development for military purposes. Several of these countries had some form of operational radar equipment in military service at the start of World War II.

The first observation of the radar effect at the U.S. Naval Research Laboratory (NRL) in Washington, D.C., was made in 1922. NRL researchers positioned a radio transmitter on one shore of the Potomac River and a receiver on the other. A ship sailing on the river unexpectedly caused fluctuations in the intensity of the received signals when it passed between the transmitter and receiver. (Today, such a configuration would be called bistatic radar.) In spite of the promising results of this experiment, U.S. Navy officials were unwilling to sponsor further work. The principle of radar was "rediscovered" at NRL in 1930 when L. A. Hyland observed that an aircraft flying through the beam of a transmitting antenna caused a fluctuation in the received signal. Although Hyland and his associates at NRL were enthusiastic about the prospect of detecting targets by radio means and were eager to pursue its development in earnest, little interest was shown by higher authorities in the navy. Not until it was learned how to use a single antenna for both transmitting and receiving (now termed monostatic radar) was the value of radar for detecting and tracking aircraft and ships fully recognized. Such a system was demonstrated at sea on the battleship USS *New York* in early 1939.

The first radars developed by the U.S. Army were the SCR-268 (at a frequency of 205 MHz) for controlling antiaircraft gunfire and the SCR-270 (at a frequency of 100 MHz) for detecting aircraft. Both of these radars were available at the start of World War II, as was the navy's CXAM shipboard surveillance radar (at a frequency of 200 MHz). It was an SCR-270, one of six available in Hawaii at the time, that detected the approach of Japanese warplanes toward Pearl Harbor, near Honolulu, on Dec. 7, 1941; however, the significance of the radar observations was not appreciated until bombs began to fall.

Britain commenced radar research for aircraft detection in 1935. By September 1938 the first British radar system, the Chain Home, had gone into 24-hour operation, and it remained operational throughout the war. The Chain

Home radars allowed Britain to deploy successfully its limited air defenses against the heavy German air attacks conducted during the early part of the war. They operated at about 30 MHz—in what is called the shortwave, or high-frequency (HF), band—which is actually quite a low frequency for radar. It might not have been the optimum solution, but the inventor of British radar, Sir Robert Watson-Watt, believed that something that worked and was available was better than an ideal solution that was only a promise or might arrive too late.

The Soviet Union also started working on radar during the 1930s. At the time of the German attack on their country in June 1941, the Soviets had developed several different types of radars and had in production an aircraft-detection radar that operated at 75 MHz (in the very-high-frequency [VHF] band). Their development and manufacture of radar equipment was disrupted by the German invasion, and the work had to be relocated.

At the beginning of World War II, Germany had progressed further in the development of radar than any other country. The Germans employed radar on the ground and in the air for defense against Allied bombers. Radar was installed on a German pocket battleship as early as 1936. Radar development was halted by the Germans in late 1940 because they believed the war was almost over. The United States and Britain, however, accelerated their efforts. By the time the Germans realized their mistake, it was too late to catch up.

Except for some German radars that operated at 375 and 560 MHz, all of the successful radar systems developed prior to the start of World War II were in the VHF band, below about 200 MHz. The use of VHF posed several problems. First, VHF beamwidths are broad. (Narrow beamwidths yield greater accuracy, better resolution, and the exclusion of unwanted echoes from the ground or other clutter.) Second, the VHF portion of the electromagnetic spectrum does not permit the wide bandwidths required for the short pulses that allow for greater accuracy in range determination. Third, VHF is subject to atmospheric noise, which limits receiver sensitivity. In spite of these drawbacks, VHF represented the frontier of radio technology in the 1930s, and radar development at this frequency range constituted a genuine pioneering accomplishment.

THE ATOMIC CLOCK

In 1913 the Danish physicist Niels Bohr postulated that atoms exist in states of discrete energy and that any transition between two energy states is accompanied by absorption or emission of a photon of a specific frequency. In an unperturbed atom (that is, an atom not affected by neighbouring atoms or external fields), the energies of the various states depend only upon intrinsic features of atomic structure, which are understood not to vary. A transition between a pair of these states, therefore, involves the absorption or emission of a photon with an invariable frequency; that frequency is designated the fundamental frequency of the transition.

In many atoms and molecules, transitions between energy states involve sharply defined frequencies in the vicinity of 1010 hertz. After dependable methods of generating such frequencies were developed during World War II for microwave radar, these transitions were applied to problems of timekeeping. In 1946 principles of the use of atomic and molecular transitions for regulating the frequency of electronic oscillators were described, and in 1947 an oscillator was constructed that was controlled by energy transitions of the ammonia molecule. An ammonia-controlled clock was built in 1949 at the National Bureau of Standards, Washington, D.C.; in this clock the frequency did not vary by more than one part in 10^8.

In 1938 the so-called resonance technique of manipulating a beam of atoms or molecules was introduced. This technique was adopted in several attempts to construct a cesium-beam atomic clock, and in 1955 the first such clock was placed in operation at the National Physical Laboratory, Teddington, Eng. Between 1955 and 1958 the National Physical Laboratory and the U.S. Naval Observatory conducted a joint experiment to determine the frequency maintained by the cesium-beam clock at Teddington in terms of the ephemeris second, as established by precise observations of the Moon from Washington, D.C. The radiation associated with the particular transition of the cesium-133 atom was found to have the fundamental frequency of 9,192,631,770 hertz (cycles per second) of Ephemeris Time. In 1967 the 13th General Conference on Weights and Measures redefined the second, the unit of time in the International System of Units, as "the duration of 9,192,631,770 periods of the radiation corresponding to the transition between the two hyperfine levels of the ground state of the cesium-133 atom."

Until the 1990s the cesium beam atomic clock was the most accurate standard of atomic time and frequency. The principle underlying the cesium clock is that all atoms of cesium-133 are identical and, when they absorb or release energy, produce radiation of exactly the same frequency, which makes the atoms perfect timepieces. Since that time, laboratories around the world have steadily improved the accuracy of cesium fountain atomic clocks. These clocks get their name from the fountainlike motion of the constituent cesium gas. The timing process begins by introducing cesium gas into a vacuum chamber and directing six infrared lasers (located at right angles to one another) to compact and cool (slow down) the cesium atoms to a temperature near absolute zero. Then two vertical lasers are used to nudge the atoms up about a metre (creating a "fountain") through a microwave-filled cavity. The microwave frequency is tuned to maximize the observed fluorescence, which occurs at the natural resonance frequency (9,192,631,770 hertz) of the cesium atom. Because the round-trip through the microwave cavity takes about a second, control of the microwave frequency has resulted in greater timekeeping accuracy. The best cesium fountain atomic clocks are now predicted to be off by less than one second in more than 50 million years.

Chapter 8: Agriculture and Industry

The growing of living things and the fabricating of new materials and objects are fundamental to the maintenance of life. They are also fundamental to defining human beings as separate from all other living creatures. From the first forging of bronze to the creation of microscopic carbon buckyballs, from the fashioning of looms to the building of the combine harvester, the innovations profiled here demonstrate the resolve of human beings to make their material lives safe, comfortable, and prosperous.

BRONZE

At some time in the 4th millennium BCE, during the period of the earliest use of copper, a new kind of "copper" happened to be made by smelting together two separate ores, one bearing copper, the other tin. The resulting metal was recognized as being far more useful than copper alone, and the short period of copper tools came to an end. The new metal, a copper–tin alloy of mostly copper, was bronze. It was produced in the fluid state at a temperature less than that needed for copper, could be formed economically by casting, and could be hammer-hardened more than copper. The tin noticeably increased the liquidity of the melt, checked the absorption of oxygen and other gases, and suppressed the formation of cuprous oxide, all features that facilitated the casting operation. A two-piece, or split, mold, impractical for copper, worked very well with bronze. Bronze also expanded just a bit before solidifying and thus picked up the detail of a mold before it contracted in cooling.

The production of bronze was an invention in its true sense, and the period of its extensive and characteristic use has been designated the Bronze Age. True bronze seems to have appeared between 3000 and 2500 BCE, beginning in the Tigris-Euphrates delta. While there may have been some independent development of bronze in varying localities, it is most likely that the bronze culture spread through trade and the migration of peoples from the Middle East to Egypt. From Egypt the use of bronze rapidly spread over the Mediterranean area: to Crete in 3000 BCE, to Sicily in 2500 BCE, to France and other parts of Europe in 2000 BCE, and to Britain and the Scandinavian area in 1800 BCE.

The earliest bronzes were of uneven composition. Later, the tin content was controlled at about 10 percent, a little less for hammered goods, a little more for ornamental castings. The edges of hammered bronze tools of this composition were more than twice as hard as those obtained from copper.

One of the most important new implements was the sword. With the sword there was for the first time in European history an object entirely dedicated to fighting and not doubling as a tool. Fighting is evident from earlier periods as well, but during the Bronze Age it was formalized. Toward the Late Bronze Age the warrior emerged, sheathed in an assemblage of defensive items: the armour. Breastplates of bronze, at first beaten and then cast to the warrior's individual shape, were commonplace among heavy infantry and elite cavalry. Greaves, defenses for the lower leg, closely followed the breastplate. At first these were forged of bronze plates; some classical Greek examples sprang open and could be snapped onto the calf. Defenses for more remote portions of the body, such as vambraces for the forearm and defenses for the ankle resembling spats, were included in Greek temple dedications, but they were probably not common in field service. Bronze was the most common metal for body defenses well into the Iron Age, a consequence of the fact that it could be worked in large pieces without extended hand forging and careful tempering, while iron had to be forged from relatively small billets.

The development of bronze, and later iron, technology led to the making of metal tools for working wood, such as axes and saws. Less effort was thus required to fell and work large trees. This led in turn to new developments in building techniques; timbers were cut and shaped extensively, hewed into square posts, sawed into planks, and split into shingles. Log cabin construction appeared in the forested areas of Europe, and timber framing became more sophisticated. Although the excavated remains are fragmentary, undoubtedly major advances were made in timber technology in this period. Some of the products, such as the sawed plank and the shingle, are still used today.

When copper and bronze were first used in Asia is not known. The epics of the Shujing mention the use of copper in China as early as 2500 BCE, but nothing is known of the state of the art at that time or of the use of the metal prior to that time. Bronze vessels of great beauty made during the Shang dynasty (1766–1122 BCE) have been found, indicating an advanced art. The source of the metals, however, is unknown.

The Copper Age in the Americas probably dawned between 100 and 200 CE. Native copper was mined and used extensively and, though some bronze appeared in South America, its use developed slowly until after the arrival of Columbus and other European explorers. Both

North and South America passed more or less directly from the Copper Age into the Iron Age.

Besides its traditional use in weapons and tools, bronze has also been widely used in coinage; most "copper" coins are actually bronze, typically with about 4 percent tin and 1 percent zinc. Other modern bronzes contain alloying metals other than tin. Bell metal, characterized by its sonorous quality when struck, is a bronze with a high tin content of 20–25 percent. Statuary bronze, with a tin content of less than 10 percent and an admixture of zinc and lead, is technically a brass. Bronze is improved in hardness and strength by the addition of a small amount of phosphorus; phosphor bronze may contain 1 or 2 percent phosphorus in the ingot and a mere trace after casting, but its strength is nonetheless enhanced for such applications as pump plungers, valves, and bushings. Also useful in mechanical engineering are manganese bronzes, in which there may be little or no tin but considerable amounts of zinc and up to 4.5 percent manganese. Aluminum bronzes, containing up to 16 percent aluminum and small amounts of other metals such as iron or nickel, are especially strong and corrosion-resistant; they are cast or wrought into pipe fittings, pumps, gears, ship propellers, and turbine blades.

IRON

There is evidence that meteorites were used as a source of iron before 3000 BCE, but extraction of the metal from ores dates from about 2000 BCE. Production seems to have started in the copper-producing regions of Anatolia and Persia, where the use of iron compounds as fluxes to assist in melting may have accidentally caused metallic iron to accumulate on the bottoms of copper smelting furnaces. When iron making was properly established, smelting involved creating a bed of red-hot charcoal to which iron ore mixed with more charcoal that was added. Chemical reduction of the ore then occurred, but since primitive furnaces were incapable of reaching temperatures higher than 2,100°F (1,150°C), the normal product was a solid lump of metal known as a bloom. This may have weighed up to 11 pounds (5 kg) and consisted of almost pure iron with some entrapped slag and pieces of charcoal. The manufacture of iron artifacts then required a shaping operation, which involved heating blooms in a fire and hammering the red-hot metal to produce the desired objects. Iron made in this way is known as wrought iron. Sometimes too much charcoal seems to have been used, and iron-carbon alloys, which have lower melting points and can be cast into simple shapes, were made unintentionally. The applications of this cast iron were limited because of its brittleness, and in the early Iron Age only the Chinese seem to have exploited it. Elsewhere, wrought iron was the preferred material.

Although the Romans built furnaces with a pit into which slag could be run off, little change in iron-making methods occurred until medieval times. By the 15th century, many bloomeries used low shaft furnaces with waterpower to drive the bellows, and the bloom, which might weigh over 250 pounds (100 kg), was extracted through the top of the shaft. The final version of this kind of bloomery hearth was the Catalan forge, which survived in Spain until the 19th century. Another design, the high bloomery furnace, had a taller shaft and evolved into the 10-foot-high (3 m) *Stückofen*, which produced blooms so large they had to be removed through a front opening in the furnace.

The blast furnace appeared in Europe in the 15th century when it was realized that cast iron could be used to make one-piece guns with good pressure-retaining properties. At first, the differences between a blast furnace and a *Stückofen* were slight. Both had square cross sections, and the main changes required for blast-furnace operation were an increase in the ratio of charcoal to ore in the charge and a taphole for the removal of liquid iron. The product of the blast furnace became known as "pig iron" from the method of casting, which involved running the liquid into a main channel connected at right angles to a number of shorter channels. The whole arrangement resembled a sow suckling her litter, and so the lengths of solid iron from the shorter channels were known as "pigs."

Despite the military demand for cast iron, most civil applications required malleable iron, which until then had been made directly in a bloomery. The arrival of blast furnaces, however, opened up an alternative manufacturing route; this involved converting cast iron to wrought iron by a process known as fining. Pieces of cast iron were placed on a finery hearth, on which charcoal was being burned with a plentiful supply of air, so that carbon in the iron was removed by oxidation, leaving semisolid malleable iron behind. From the 15th century on, this two-stage process gradually replaced direct iron making, which nevertheless survived into the 19th century.

By the middle of the 16th century, blast furnaces were being operated more or less continuously in southeastern England. Increased iron production led to a scarcity of wood for charcoal and to its subsequent replacement by coal in the form of coke—a discovery that is usually credited to Abraham Darby in 1709. Because the higher strength of coke enabled it to support a bigger charge, much larger furnaces became possible, and weekly outputs of 5 to 10 tons of pig iron were achieved.

Next, the advent of the steam engine to drive blowing cylinders meant that the blast furnace could be provided with more air. This created the potential problem that pig iron production would far exceed the capacity of the finery process. Accelerating the conversion of pig iron to malleable iron was attempted by a number of inventors, but the most successful was the Englishman Henry Cort, who patented his puddling furnace in 1784. Cort used a coal-fired reverberatory furnace to melt a charge of pig iron to which iron oxide was added to make a slag. Agitating the resultant "puddle" of metal caused carbon to be removed by oxidation (together with silicon, phosphorus, and manganese). As a result, the melting point of the metal rose so that it became semisolid, although the slag remained quite fluid. The metal was then formed into balls and freed from as much slag as possible before being removed from the furnace and squeezed in a hammer. For a short time, puddling furnaces were able to provide enough iron to meet the demands for machinery. Soon, though, blast-furnace capacity raced ahead as a result of the Scotsman James Beaumont Nielsen's invention in 1828 of the hot-blast stove for preheating blast air and the realization that a round furnace performed better than a square one.

The eventual decline in the use of wrought iron was brought about by a series of inventions that allowed furnaces to operate at temperatures high enough to melt iron. It was then possible to produce steel.

THE PLOWSHARE

The antecedent of the plow is the prehistoric digging stick. The earliest plows were doubtless digging sticks fashioned with handles for pulling or pushing. Iron Age technology was applied to agriculture during Roman times in the form of light, wheelless iron (or iron-tipped) plowshares drawn by oxen, which opened up the possibility of

A tractor pulling a disk plow. Courtesy of John Deere

deeper plowing and of cultivating heavier soils than those normally worked in the Greco-Roman period. The construction of plows improved slowly during these centuries, but the moldboard for turning over the earth did not appear until the 11th century CE, so that the capacity of turning the sod depended more on the wrists of the plowman than on the strength of his draft team. This discouraged tackling heavy ground. The potentialities of the heavy plow were thus not fully exploited in the temperate areas of Europe until after the Roman period.

Though the Roman historian Pliny the Elder claimed a wheeled plow was used in Cisalpine Gaul (northern Italy) about the time of Christ, there is a good deal of doubt about that. A wheeled asymmetrical plow was certainly in use in some parts of western Europe by the late 10th century. Illuminated manuscripts and somewhat later calendars show a plow with two wheels fitted with a rudimentary moldboard and a heavy knife, or coulter, to dig under the surface. Whereas earlier plows had merely scratched the surface of the soil, this plow could invert the soil and turn a true furrow, thus making a better seedbed. Its use left high ridges on the land, traces of which can still be seen in some places. Yet because the new plow required a team of eight oxen—more than any single peasant owned—plowing was pooled. Such a system allowed little room for individual initiative; everyone followed established routines, with the pace of the work set by the ox team.

The horse collar, which replaced the old harness band that pressed upon the animal's windpipe, severely restricting its tractive power, was one of the most important inventions in the history of agriculture. Apparently invented in China, the rigid, padded horse collar allowed the animal to exert its full strength, enabling it to do heavier work, plowing as well as haulage. Many peasants continued to use oxen, however, because horses were more

A tractor pulling a large chisel plow. Courtesy of John Deere

expensive to buy and to keep. Some plowing was done by two oxen as in former times; four, eight, or more were occasionally necessary for very difficult land.

The 18th-century addition of the moldboard, which turned the furrow slice cut by the plowshare, was another important advance. In the mid-19th century the black prairie soils of the American Midwest challenged the strength of the existing plow, and American mechanic John Deere invented the all-steel one-piece share and moldboard. The three-wheel sulky plow followed and, with the introduction of the gasoline engine, the tractor-drawn plow.

In its simplest form the moldboard plow consists of the share, the broad blade that cuts through the soil; the moldboard, for turning the furrow slice; and the landside, a plate on the opposite side from the moldboard that absorbs the side thrust of the turning action. Horse-drawn moldboard plows had a single bottom (share and moldboard), while tractor-drawn plows have from one to 14 hydraulically lifted and controlled bottoms staggered in tandem. Listers and middlebusters are double-moldboard plows that leave a furrow by throwing the dirt both ways.

Disk plows usually have three or more individually mounted concave disks that are inclined backward to achieve maximum depth. They are particularly adapted for use in hard, dry soils, shrubby or bushy land, or on rocky land. Disk tillers, also called harrow plows or one-way disk plows, usually consist of a gang of many disks mounted on one axle. Used after grain harvest, they usually leave some stubble to help reduce wind erosion and often have seeding equipment. Two-way (reversible) plows have disks or moldboards that can be either opposed, so that one fills the trench made by the other, or set to throw the soil entirely to the right or left.

THE POWER LOOM

Woven cloth is normally much longer in one direction than the other. The lengthwise threads are called the warp. The other threads, which are combined with the warp and lie widthwise, are called the weft. An individual thread from the warp, of indefinite length, is called an end. Each individual length of weft, extending from one edge of the cloth to the other, is called a pick, or shot. Consecutive picks are usually consecutive lengths of one piece of weft yarn that is repeatedly folded back on itself.

The Basic Weaving Process

In all methods of weaving cloth, before a length of weft is inserted in the warp, the warp is separated—over a short length extending from the cloth already formed—into two sheets. The process is called shedding, and the space between the sheets is called the shed. A pick of weft is then laid between the two sheets of warp, in the operation known as "picking." A new shed is then formed in accordance with the desired weave structure, with some or all of the ends in each sheet moving over to the position previously occupied by the other sheet. In this way the weft is clasped between two layers of warp.

Since it is not possible to lay the weft close to the junction of the warp and the cloth already woven, a further operation called "beating in," or "beating up," is necessary to push the pick to the desired distance away from the last one inserted previously. Although beating in usually takes place while the shed is changing, it is normally completed before the new shed is fully formed.

The sequence of primary operations in one weaving cycle is thus shedding, picking, and beating in. Any set of devices permitting a warp to be tensioned and a shed to be formed is called a loom (from Middle English *lome*, "tool"). In most looms, the warp shed is formed with the aid of devices called heddles, and the weft is supplied from a shuttle, a hollow projectile inside which a weft package is mounted in such a way that the weft can be freely unwound through an eyelet leading from the inside to the outside. The shuttle enters the shed and traverses the warp, leaving a trail of weft behind.

AUTOMATED AND POWER-DRIVEN LOOMS

The first decisive step toward automation of the loom was the invention of the flying shuttle patented in 1733 by the Englishman John Kay, a weaver of broadloom fabrics. Because of their width, broadloom fabrics required two weavers to sit side by side, one throwing the shuttle from the right to the centre, the other reaching between the warps and sending it on its way to the left and then returning it to the centre. The stopping of the shuttle and the reaching between the warps caused imperfections in the cloth. Kay devised a mechanical attachment controlled by a cord jerked by the weaver that sent the shuttle flying through the shed. Jerking the cord in the opposite direction sent the shuttle on its return trip. Using the flying shuttle, one weaver could weave fabrics of any width more quickly than two could before. A more important virtue of Kay's invention, however, lay in its adaptability to automatic weaving.

The first power-driven machine for weaving fabric-width goods was patented in 1785 by Edmund Cartwright, an English clergyman. It was inadequate because it considered only three motions: shedding, picking, and winding the woven cloth onto the cloth beam. Cartwright's second patent (1786) proved too ambitious, but his concept of a weaving machine became the basis for the successful power loom.

One of the great obstacles to the success of the power loom was the necessity to stop the loom frequently in order to dress (i.e., apply sizing to) the warp. This was an operation that, like many others, had been done in proportionately reasonable time when the weaving was done by hand. With the power loom a second man had to be employed continuously to do this work, so there was no saving of expense or time. In the early 19th century a dressing machine was developed that prepared the warp after it had been wound onto the warp beam and as it was passed to the cloth beam. Although later superseded by an improved sizing apparatus, this device made the power loom a practical tool. Modern looms still weave by repeating in sequence the operations of shedding, picking, and beating in.

CANNING

Canning is a method of preserving food from spoilage by storing it in containers that are hermetically sealed and then sterilized by heat. The process was invented after prolonged research by Nicolas Appert of France. Inspired by the French Directory's offer of a prize for a way to conserve food for transport, Appert began a 14-year period of experimentation in 1795. Using corked-glass containers reinforced with wire and sealing wax and kept in boiling water for varying lengths of time, he preserved soups, fruits, vegetables, juices, dairy products, marmalades, jellies, and syrups. A 12,000-franc award in 1810 specified that he publish his findings, which appeared that year as *L'Art de conserver, pendant plusieurs années, toutes les substances animales et végétales* (*The Art of Preserving All Kinds of Animal and*

Vegetable Substances for Several Years). He used the money to establish the first commercial cannery, the House of Appert, at Massy, which operated from 1812 until 1933.

It was 50 years before Louis Pasteur was able to explain why the food so treated did not spoil: the heat killed the microorganisms in the food, and the sealing kept other microorganisms from entering the jar. In 1810 Peter Durand of England patented the use of tin-coated iron cans instead of bottles, and by 1820 he was supplying canned food to the Royal Navy in large quantities. European canning methods reached the United States soon thereafter, and that country eventually became the world leader in both automated canning processes and total can production. In the late 19th century, Samuel C. Prescott and William Underwood of the United States set canning on a scientific basis by describing specific time-temperature heating requirements for sterilizing canned foods.

Originally, cans consisted of a sheet of tin-plated iron that was rolled into a cylinder (known as the body), onto which the top and bottom were manually soldered. This form was replaced in the early 20th century by the modern sanitary, or open-top, can, whose constituent parts are joined by interlocking folds that are crimped, or pressed together. Polymer sealing compounds are applied to the end, or lid, seams, and the body seams can be sealed on the outside by soldering. The modern tin can is made of 98.5 percent sheet steel with a thin coating of tin (i.e., tinplate). It is manufactured on wholly automatic lines of machinery at rates of hundreds of cans per minute.

Most vegetables, fruits, meat and dairy products, and processed foods are stored in tin cans, but soft drinks and many other beverages are now commonly stored in aluminum cans, which are lighter and do not rust. Aluminum cans are made by impact extrusion; the body of the can is punched out in one piece from a single aluminum sheet by a stamping die. This seamless piece, which has a rounded bottom, is then capped with a second piece as its lid. The tabs used in pop-top cans are also made of aluminum. Bimetal cans are made of aluminum bodies and steel lids.

REFRIGERATION

Before mechanical refrigeration systems were introduced, ancient peoples, including the Greeks and Romans, cooled their food with ice transported from the mountains. Wealthy families made use of snow cellars, pits that were dug into the ground and insulated with wood and straw, to store the ice. In this manner, packed snow and ice could be preserved for months. Stored ice was the principal means of refrigeration until the beginning of the 20th century, and it is still used in some areas.

In India and Egypt evaporative cooling was employed. If a liquid is rapidly vaporized, it expands quickly. The rising molecules of vapour abruptly increase their kinetic energy. Much of this increase is drawn from the immediate surroundings of the vapour, which are therefore cooled. Thus, if water is placed in shallow trays during the cool tropical nights, its rapid evaporation can cause ice to form in the trays, even if the air does not fall below freezing temperatures. By controlling the conditions of evaporation, it is possible to form even large blocks of ice in this manner.

Cooling caused by the rapid expansion of gases is the primary means of refrigeration today. The technique of evaporative cooling, as described heretofore, has been known for centuries, but the fundamental methods of mechanical refrigeration were only discovered in the middle of the 19th century.

The first known artificial refrigeration was demonstrated by William Cullen at the University of Glasgow in

1748. Cullen let ethyl ether boil into a partial vacuum; he did not, however, use the result to any practical purpose. In 1805 an American inventor, Oliver Evans, designed the first refrigeration machine that used vapour instead of liquid. Evans never constructed his machine, but one similar to it was built by an American physician, John Gorrie, in 1844.

Commercial refrigeration is believed to have been initiated by an American businessman, Alexander C. Twinning, in 1856. Shortly afterward, an Australian, James Harrison, examined the refrigerators used by Gorrie and Twinning and introduced vapour-compression refrigeration to the brewing and meat-packing industries. A somewhat more complex system was developed by Ferdinand Carré of France in 1859. Unlike earlier vapour-compression machines, which used air as a coolant, Carré's equipment contained rapidly expanding ammonia. (Ammonia liquefies at a much lower temperature than water and is thus able to absorb more heat.) Carré's refrigerators were widely used, and vapour-compression refrigeration became, and still is, the most widely used method of cooling.

The basic components of a modern vapour-compression refrigeration system are a compressor; a condenser; an expansion device, which can be a valve, a capillary tube, an engine, or a turbine; and an evaporator. The gas coolant is first compressed, usually by a piston, and then pushed through a tube into the condenser. In the condenser, the winding tube containing the vapour is passed through either circulating air or a bath of water, which removes some of the heat energy of the compressed gas. The cooled vapour is passed through an expansion valve to an area of much lower pressure; as the vapour expands, it draws the energy of its expansion from its surroundings or the medium in contact with it. Evaporators may directly cool a space by letting the vapour come into contact with the area to be chilled, or they may act indirectly—i.e., by cooling a secondary medium such as water. In most domestic refrigerators, the coil containing the evaporator directly contacts the air in the food compartment. At the end of the process, the hot gas is drawn toward the compressor.

In spite of the successful early use of ammonia as the coolant, that substance had a severe disadvantage: if it leaked, it was unpleasant as well as toxic. Refrigeration engineers searched for acceptable substitutes until the 1920s, when a number of synthetic refrigerants were developed. The best known of these substances was patented under the brand name of Freon. Chemically, Freon was created by the substitution of two chlorine and two fluorine atoms for the four hydrogen atoms in methane (CH_4); the result, dichlorofluoromethane (CCl_2F_2), is odourless and is toxic only in extremely large doses. However, in the mid-1970s, photochemical dissociation of Freons and related chlorofluorocarbons (CFCs) was implicated as a major cause of the apparent degradation of the Earth's ozone layer. Depletion of the ozone could create a threat to animal life on the Earth because ozone absorbs ultraviolet radiation that can induce skin cancer. The use of Freons in aerosol-spray containers was banned in the United States in the late 1970s. By the early 1990s, accumulating evidence of ozone depletion in the polar regions had heightened worldwide public alarm over the problem, and by 1996 most of the developed nations had banned production of nearly all Freons.

STEEL

The most important development of the 19th century was the large-scale production of cheap steel. As mentioned previously, prior to about 1850, wrought iron was produced by a process known as puddling and steel by the technique of crucible melting. In the puddling process, invented in Great Britain in 1784, hot gases were drawn over a charge of pig iron and iron ore held on the furnace hearth. The melted product was stirred with iron rabbles (rakes), and, as it became pasty with loss of carbon, it was worked into balls, which were subsequently forged or rolled to a useful shape. In the crucible process, introduced in England in 1740, bar iron and added materials were placed in clay crucibles heated by coke fires, which resulted in a reliable steel.

Both puddling and crucible melting were conducted in small-scale units without significant mechanization. The first change was the development of the open-hearth furnace by William and Friedrich Siemens in Britain and by Pierre and Émile Martin in France. Employing the regenerative principle, in which outgoing combusted gases are used to heat the next cycle of fuel gas and air, this enabled high temperatures to be achieved while saving on fuel. Pig iron could then be taken through to molten iron or low-carbon steel without solidification, scrap could be added and melted, and iron ore could be melted into the slag above the metal to give a relatively rapid oxidation of carbon and silicon—all on a much enlarged scale. Another major advance was Henry Bessemer's process, patented in 1855 and first operated in 1856, in which air was blown through molten pig iron from tuyeres set into the bottom of a pear-shaped vessel called a converter. Heat released by the oxidation of dissolved silicon, manganese, and carbon was enough to raise the temperature above the melting point of the refined metal (which rose as the carbon content was lowered) and thereby maintain it in the liquid state. Very soon Bessemer had tilting converters producing 5 tons in a heat of one hour, compared with four to six hours for 110 pounds (50 kilograms) of crucible steel and two hours for 550 pounds (250 kilograms) of puddled iron.

THE BESSEMER CONVERTER

The first method discovered for mass-producing steel was named after its inventor, Sir Henry Bessemer of England, although the process was apparently conceived independently and almost concurrently by Bessemer and by William Kelly of the United States. As early as 1847, Kelly, a businessman-scientist of Pittsburgh, Pa., began experiments aimed at developing a revolutionary means of removing impurities from pig iron by an air blast. Kelly theorized that not only would the air, injected into the molten iron, supply oxygen to react with the impurities, converting them into oxides separable as slag, but that the heat evolved in these reactions would increase the temperature of the mass, keeping it from solidifying during the operation. After several failures, he succeeded in proving his theory and rapidly producing steel ingots.

In 1856 Bessemer, working independently in Sheffield, developed and patented the same process. Whereas Kelly had been unable to perfect the process owing to a lack of financial resources, Bessemer was able to develop it into a commercial success. The Bessemer converter was a cylindrical steel pot approximately 20 feet (6 m) high, originally lined with a siliceous refractory. Air was blown in through openings (tuyeres) near the bottom, creating oxides of silicon and manganese, which became part of the slag, and of carbon, which were carried out in the stream of air. Within a few minutes an ingot of steel could be produced, ready for the forge or rolling mill.

The original Bessemer converter was not effective in removing the phosphorus present in sizable amounts in most British and European iron ore. The invention in England, by Sidney Gilchrist Thomas, of what is now called the Thomas-Gilchrist converter, overcame this problem. The converter was lined with a basic material such as burned limestone rather than an (acid) siliceous material. Another drawback to Bessemer steel was the retention of a small percentage of nitrogen from the air blow. The open-hearth process, which was developed in the 1860s, did not suffer from this difficulty. It eventually outstripped the Bessemer process to become the dominant steelmaking process until the mid-20th century, when it was in turn replaced by the basic oxygen process.

Neither the open-hearth furnace nor the Bessemer converter could remove phosphorus from the metal, so low-phosphorus raw materials had to be used. This restricted their use from areas where phosphoric ores, such as those of the Minette range in Lorraine, were a main European source of iron. The problem was solved by Sidney Gilchrist Thomas, who demonstrated in 1876 that a basic furnace lining consisting of calcined dolomite, instead of an acidic lining of siliceous materials, made it possible to use a high-lime slag to dissolve the phosphates formed by the oxidation of phosphorus in the pig iron. This principle was eventually applied to both open-hearth furnaces and Bessemer converters.

As steel was now available at a fraction of its former cost, it saw an enormously increased use for engineering and construction. Soon after the end of the century it replaced wrought iron in virtually every field. Then, with the availability of electric power, electric-arc furnaces were introduced for making special and high-alloy steels. The next significant stage was the introduction of cheap oxygen, made possible by the invention of the Linde-Frankel cycle for the liquefaction and fractional distillation of air. The Linz-Donawitz process, invented in Austria shortly after World War II, used oxygen supplied as a gas from a tonnage oxygen plant, blowing it at supersonic velocity into the top of the molten iron in a converter vessel. As the ultimate development of the Bessemer/Thomas process, oxygen blowing became universally employed in bulk steel production.

ALUMINUM

Before 5000 BCE people in Mesopotamia were making fine pottery from a clay that consisted largely of an aluminum compound, and almost 4,000 years ago Egyptians and Babylonians used aluminum compounds in various chemicals and medicines. Pliny refers to "alumen," known now as alum, a compound of aluminum widely employed in the ancient and medieval world to fix dyes in textiles. By the 18th century, the earthy base alumina was recognized as the potential source of a metal.

EARLY LABORATORY EXTRACTION

The English chemist Humphry Davy in 1807 attempted to extract the metal. Though unsuccessful, he satisfied himself that alumina had a metallic base, which he named "alumium" and later changed to "aluminum." The name has been retained in the United States but modified to "aluminium" in many other countries.

Danish physicist and chemist Hans Christian Ørsted finally produced aluminum in 1825. "It forms," Ørsted reported, "a lump of metal which in color and luster somewhat resembles tin." A few years later Friedrich Wöhler, a German chemist at the University of Göttingen, made metallic aluminum in particles as large as pinheads and first determined the following properties of aluminum: specific gravity, ductility, colour, and stability in air.

Aluminum remained a laboratory curiosity until a French scientist, Henri Sainte-Claire Deville, announced a major improvement in Wöhler's method, which permitted Wöhler's "pinheads" to coalesce into lumps the size of marbles. Deville's process became the foundation of the aluminum industry. Bars of aluminum, made at Javel Chemical Works and exhibited in 1855 at the Paris Exposition Universelle, introduced the new metal to the public.

Electrolytic Production

Although enough was then known about the properties of aluminum to indicate a promising future, the cost of the chemical process for producing the metal was too high to permit widespread use. But important improvements presently brought breakthroughs on two fronts: first, the Deville process was improved; and, second, the development of the dynamo made available a large power source for electrolysis, which proved highly successful in separating the metal from its compounds.

The modern electrolytic method of producing aluminum was discovered almost simultaneously, and completely independently, by Charles M. Hall of the United States and Paul-Louis-Toussaint Héroult of France in 1886. The essentials of the Hall-Héroult processes were identical and remain the basis for today's aluminum industry. Purified alumina is dissolved in molten cryolite and electrolyzed with direct current. Under the influence of the current, the oxygen of the alumina is deposited on the carbon anode and is released as carbon dioxide, while free molten aluminum—which is heavier than the electrolyte—is deposited on the carbon lining at the bottom of the cell.

Hall immediately recognized the value of his discovery. He applied July 9, 1886, for a U.S. patent and worked energetically at developing the process. Héroult, on the other hand, although he applied several months earlier for patents, apparently failed to grasp the significance of the process. He continued work on a second successful process that produced an aluminum-copper alloy. Conveniently, in 1888, an Austrian chemist, Karl Joseph Bayer, discovered an improved method for making pure alumina from low-silica bauxite ores.

Hall and a group of businessmen established the Pittsburgh Reduction Company in 1888 in Pittsburgh. The first ingot was poured in November that year. Demand for aluminum grew, and a larger reduction plant was built at New Kensington, Pa., using steam-generated electricity to produce one ton of aluminum per day by 1894. The need for cheap, plentiful hydroelectric power led the young company to Niagara Falls, where in 1895 it became the first customer for the new Niagara Falls power development. In a short time, the demand for aluminum exceeded Hall's most optimistic expectations. In 1907 the company changed its name to Aluminum Company of America (Alcoa). Until World War II it remained the sole U.S. producer of primary aluminum.

Neuhausen, Switz., is the "nursery" of the European aluminum industry. There, to take advantage of waterpower available from the falls of the Rhine, Héroult built his first aluminum-bronze production facility, which later became the Aluminium-Industrie-Aktien-Gesellschaft.

SHEET AND PLATE GLASS

The Romans were perhaps the first to develop flat glass for use as windows: a bathhouse window of greenish blue colour, most likely obtained by casting, was discovered in the ruins of Pompeii. In the Middle Ages the crown process for making window glass was developed by the Normans. A mass of glass was gathered and blown into a globe at the end of the blowing iron and marvered to a conical shape. A pontil rod was attached to the other end, and the blowing iron was cracked off, leaving a jagged opening. The glassmaker then took the globe into the "glory hole" (the mouth) of the furnace, reheating it and at the same time spinning it to keep it from sagging. At some point, centrifugal force caused the globe to "flash" into a flat disk, which grew larger with continued spinning. Upon cooling, the disk was cracked off the pontil rod. Such glass was not truly flat. The disk was very uneven, being thickest near the centre and marked by concentric circular waves; at the very middle was the fractured nub, or crown, marking the point of former attachment to the pontil. Disks more than about 5 feet (1.5 m) in diameter were hardly practical.

Most medieval church windows were made from broad glass. In this process, which continued to be practiced with variations into the 20th century, a large cylinder, as much as 20 inches (50 cm) in diameter and 70 inches (175 cm) long, was made by repeated gathering, blowing, and swinging. The cylinder was slit when cold and gradually opened with moderate reheating to become flat. Glass made from this process was flatter than crown glass and did not have the telltale crown in the middle; moreover, it could be made in much larger pieces. The use of compressed air in the early 1900s allowed the cylinders to be blown as large as 30 inches (75 cm) in diameter and up to 30 feet (9 m) in length.

Despite its advantages over crown glass, broad glass had surface waviness and variations in thickness. For a higher degree of flatness, glass had to be cast (generally on a steel table) and rolled. Two continuous machines for doing this were introduced about the turn of the 20th century: the updraw machine, designed by Émile Fourcault of Belgium; and the Irving Colburn machine, developed at the Libbey-Owens Glass Company in Charleston, W.Va., U.S. In the Fourcault process, a 3- to 6-foot (1- to 2-m) wide steel mesh bait was introduced into molten glass at the working end of the furnace. The cooled glass adhered to the bait and was pulled upward between water-cooled tubes that solidified the sheet edges. The sheet was then gripped at the edges by nonchilling asbestos rollers and pulled farther up the draw tower. The Colburn machine borrowed its design from the papermaking process. The sheet was drawn vertically from the glass surface, but, after rising only a few metres, it was gradually bent over a polished nickel-alloy roller to become horizontal, ultimately traveling into the annealing lehr.

In both the Fourcault and the Colburn processes, glass was marked with undulations caused by the pulling and rolling gear. In addition, the glass sheets, like all flat glasses produced by earlier processes, had to be ground and polished for optical clarity. The development of the twin grinding and polishing machine in 1935 at the Pilkington Brothers works in Doncaster, Eng., made it possible for plate to be made by horizontal flow through a double-roller process and then ground and polished on-line.

Finally, after seven years of intense development, Alastair Pilkington introduced in 1959 the float glass process, which altogether eliminated the need for grinding and polishing. In the float process, molten glass is brought over the lip of a broad spout, allowed to pass between rollers, and floated over a bath of molten tin in a steel container. Glass enters the container at a temperature greater than 1,800 °F (1,000 °C). It is cooled over the length of the tin bath, which has a melting point of 450°F (232°C), and exits in a nearly solidified sheet form. Under such conditions glass spreads by gravity to a thickness of 0.28 inch (7 mm), but, if it is compressed with graphite paddles or stretched with knurled rollers, glass may be made in thicknesses of 0.08 to 1 inch (2 to 25 mm) and in widths up to 13 feet (4 m).

RAYON

Rayon, the first man-made fibre, was developed in the late 19th century as a substitute for silk. It is composed of regenerated and purified cellulose derived from plant sources. Rayon is described as a regenerated fibre because the cellulose, obtained from soft woods or from the short fibres (linters) adhering to cottonseeds, is converted to a liquid compound, squeezed through tiny holes in a device called a spinnerette, and then converted back to cellulose in the form of fibre.

The first practical steps toward producing such a fibre were represented by attempts to work with the highly flammable compound nitrocellulose, produced by treating cotton cellulose with nitric acid. In 1884 and 1885 in London, British chemist Sir Joseph Wilson Swan exhibited fibres made of nitrocellulose that had been treated with chemicals in order to change the material back to nonflammable cellulose. Swan did not follow up the demonstrations of his invention, so that the development of rayon as a practical fibre really began in France, with the work of industrial chemist Hilaire Bernigaud, comte de Chardonnet, who is frequently called the "father of the rayon industry." In 1889 Chardonnet exhibited fibres made by squeezing a nitrocellulose solution through spinnerettes, hardening the emerging jets in warm air, and then reconverting them to cellulose by chemical treatment. Manufacture of "Chardonnet silk," an early type of rayon and the first commercially produced man-made fibre, began in 1891 at a factory in Besançon.

Although Chardonnet's process was simple and involved a minimum of waste, it was slow, expensive, and potentially dangerous. In 1890 another French chemist, Louis-Henri Despeissis, patented a process for making fibres from cuprammonium rayon. This material was based on the Swiss chemist Matthias Eduard Schweizer's discovery in 1857 that cellulose could be dissolved in a solution of copper salts and ammonia and, after extrusion, be regenerated in a coagulating bath. In 1908 the German textile firm J.-P. Bemberg began to produce cuprammonium rayon as Bemberg silk.

A third type of cellulose—and the type most commonly made today—was produced in 1891 from a syrupy, yellow, sulfurous-smelling liquid that three British chemists—Charles F. Cross, Edward J. Bevan, and Clayton Beadle—discovered by dissolving cellulose xanthate in dilute sodium hydroxide. By 1905 the British silk firm Samuel Courtauld & Company was producing this fibre, which became known as viscose rayon (or simply viscose). In 1911 the American Viscose Corporation began production in the United States.

Modern manufacture of viscose rayon has not changed in its essentials. Purified cellulose is first treated with caustic soda (sodium hydroxide). After the alkali cellulose has aged, carbon disulfide is added to form cellulose xanthate, which is dissolved in sodium hydroxide. This viscous solution (viscose) is forced through spinnerettes. Emerging from the holes, the jets enter a coagulating bath of acids and salts, in which they are reconverted to cellulose and coagulated to form a solid filament. The filament may be manipulated and modified during the manufacturing process to control lustre, strength, elongation, filament size, and cross section as demanded.

Rayon has many properties similar to cotton and can also be made to resemble silk. Readily penetrated by water, the fibre swells and loses strength when wet. It can be washed in mild alkaline solutions but loses strength if subjected to harsh alkalies. Common dry-cleaning solvents are not harmful. In apparel, rayon is used alone or in blends with other fibres in applications where cotton is normally used. High-strength rayon, produced by drawing (stretching) the filaments during manufacture to induce crystallization of the cellulose polymers, is made into tirecord for use in automobile tires. Rayon is also blended with wood pulp in paper making.

Rayon fibre remains an important fibre, although production has declined in industrial countries because of environmental concerns connected with the release of carbon disulfide into the air and salt by-products into streams. Such concerns have led to the development of new types of rayon such as lyocell. Lyocell is produced by dissolving wood cellulose in a nontoxic amine oxide solvent, which is washed from the regenerated fibres and recovered for reuse.

BAKELITE

The beginning of the modern plastics industry is often dated to Belgian-born American chemist Leo Hendrik Baekeland's first patent application in 1907 for Bakelite and to the founding of his General Bakelite Company in 1910. Bakelite, a hard, infusible, and chemically resistant plastic, was based on a chemical combination of phenol and formaldehyde, two compounds that were derived from coal tar and wood alcohol (methanol), respectively, at that time. This made it the first truly synthetic resin, representing a significant advance over earlier plastics that were based on modified natural materials. Because of its excellent insulating properties, Bakelite was also the first commercially produced synthetic resin, replacing shellac and hard rubber in parts for the electric power industry as well as home appliances. In the 1920s it was widely used in knobs, dials, circuitry panels, and even cabinets for radios, and it was also employed in the electrical systems of automobiles. In the 1930s cast Bakelite, along with many other competing phenolic resins, enjoyed a vogue in colourful costume jewelry and novelties.

Experiments with phenolic resins had actually preceded Baekeland's work, beginning in 1872 with the work of German chemist Adolf von Baeyer, but these trials had succeeded only in producing viscous liquids or brittle solids of no apparent value. It was Baekeland who succeeded in controlling the phenol-formaldehyde condensation reaction to produce the first synthetic resin. Baekeland was able to stop the reaction while the resin was still in a liquid state (which he called the A stage). The A resin (resol) could be made directly into a usable plastic, or it could be brought to a solid B stage (resitol), where, though almost infusible and insoluble, it could still be ground into powder and then softened by heat to final shape in a mold. Both A and B could be brought to a completely cured, thermoset C stage (Bakelite C, or true Bakelite) by being heated under pressure.

In 1909 Baekeland made the first public announcement of his invention, in a lecture before the New York section of the American Chemical Society. By 1910, Baekland had a semicommercial production operation established in his laboratory, which led him to form his

LEO BAEKELAND

Leo Hendrik Baekeland, the industrial chemist who helped found the modern plastics industry through his invention of Bakelite, was born on Nov. 14, 1863, in Ghent, Belg. He received his doctorate maxima cum laude from the University of Ghent at the age of 21 and taught there until 1889, when he went to the United States and joined a photographic firm. He soon set up his own company to manufacture his invention, Velox, a photographic paper that could be developed under artificial light. Velox was the first commercially successful photographic paper. In 1899 Baekeland sold his company and rights to the paper to the U.S. inventor George Eastman for $1,000,000.

Baekeland's search, begun in 1905, for a synthetic substitute for shellac led to the discovery in 1907 of Bakelite, a condensation product of formaldehyde and phenol that is produced at high temperature and pressure. Though the material had been reported earlier, Baekeland was the first to find a method of forming it into the thermosetting plastic. Baekeland received many honours for his invention and served as president of the American Chemical Society in 1924. He died on Feb. 23, 1944, in Beacon, N.Y.

company, and in 1911 General Bakelite began operations in Perth Amboy, N.J. In a plastics market virtually monopolized by celluloid, a highly flammable material that dissolved readily and softened with heat, Bakelite found ready acceptance because it could be made insoluble and infusible. Moreover, the resin would tolerate considerable amounts of inert ingredients and therefore could be modified through the incorporation of various fillers. For general molded parts, wood flour was preferred, but, where heat resistance, impact strength, or electrical properties were involved, other fillers such as cotton flock, asbestos, and chopped fabric were used. For the making of laminated structures, sheets of paper or fabric were impregnated with the resin in alcohol solution and then heated under pressure to form tough, rigid assemblies. Owing to the inclusion of fillers and reinforcement, Bakelite products were almost always opaque and dark-coloured.

In 1927 the Bakelite patent expired. In the growing consumer market of the 1930s and after, Bakelite also faced competition from other thermosetting resins such as urea formaldehyde and melamine formaldehyde and from new thermoplastic resins such as cellulose acetate, polyvinyl chloride, polymethyl methacrylate, and polystyrene. These new plastics could be used to produce household products in virtually any hue and in varying degrees of clarity. In 1939 Baekeland sold the Bakelite trademark to the Union Carbide and Carbon Corporation (now Union Carbide Corporation). Union Carbide sold the trademark in 1992 to the Georgia-Pacific Corporation, which employed Bakelite as a bonding agent for plywood and particleboard. Bakelite is still commonly used for dominoes, mah-jongg tiles, checkers, and chess pieces.

THE COMBINE HARVESTER

The combine harvester, employed in developed countries for the harvesting of wheat and other cereals, is the modern mechanized equivalent of the ancient scythe, sickle, and flail. The mechanical ancestor of today's large combines was the McCormick reaper, introduced in 1831 and followed by self-raking reapers that delivered the cut grain in bunches on the ground to be bound by hand. In 1843 a "stripper" was brought out in Australia that removed the wheat heads from the plants and threshed them in a single operation. Threshing machines were powered first by men or animals, often using treadmills, later by steam and internal-combustion engines. An early primitive combine was a horse-drawn "combination harvester–thresher" introduced in Michigan in 1836 and later used in California. Modern combines were not generally adopted until the 1930s, when tractor-drawn models became available. In 1940 self-propelled combines capable of cutting swaths 8 to 18 feet (2.5 to 5.5 m) wide were introduced. Self-propelled combines have brought a striking reduction in harvesting time and labour; in 1829 harvesting 1 acre of wheat required 14 man-hours, while the modern combine requires less than 30 minutes. In the early part of the 19th century harvesting a bushel of wheat required three man-hours' work; today it takes five minutes.

In design, the combine is essentially a binder-type cutting device that delivers the grain to a threshing machine modified to work as it moves across the field. The cutting–gathering component, designed to take the grain with a minimum of straw, is sometimes called the "header." A threshing cylinder rubs grain out of the heads against a concave surface. Some grain and chaff go with the straw to the straw deck, on which grain is shaken out and delivered to the cleaning shoe. Some of the grain and chaff goes directly to the cleaning shoe, on which sieves and a blast of air are used to separate and clean the grain. After passing through the air blast, the grain drops into a clean-grain auger that conveys it to an elevator and into a storage tank. Straw drops out of the back of the combine in a windrow for baling or is scattered over the ground by a fanlike spreader. Some combines for use on steeply rolling land have a body supported in a frame by hydraulic cylinders that automatically adjust to keep the body level.

INDUSTRIAL ROBOTS

Though not humanoid in form, machines with flexible behaviour and a few humanlike physical attributes have been developed for industry. The first stationary industrial robot was the programmable Unimate, an electronically controlled hydraulic heavy-lifting arm that could repeat arbitrary sequences of motions. It was invented in 1954 by the American engineer George Devol and was developed by Unimation, Inc., a company founded in 1956 by American engineer Joseph Engelberger. In 1959 a prototype of the Unimate was introduced in a General Motors Corporation (GM) die-casting factory in Trenton, N.J. In 1961 Condec Corp. (after purchasing Unimation the preceding year) delivered the world's first production-line robot to the GM factory; it had the unsavoury task (for humans) of removing and stacking hot metal parts from a die-casting machine. Unimate arms continue to be developed and sold by licensees around the world, with the automobile industry remaining the largest buyer.

More advanced computer-controlled electric arms guided by sensors were developed in the late 1960s and 1970s at the Massachusetts Institute of Technology (MIT) and at Stanford University, where they were used with cameras in robotic hand-eye research. Stanford's Victor Scheinman, working with Unimation for GM, designed the first such arm used in industry. Called PUMA (Programmable Universal Machine for Assembly), they have been used since 1978 to assemble automobile subcomponents such as dash panels and lights. PUMA was widely imitated, and its descendants, large and small, are still used for light assembly in electronics and other industries. Since the 1990s small electric arms have become important in molecular biology laboratories, precisely handling test-tube arrays and pipetting intricate sequences of reagents.

Mobile industrial robots also first appeared in 1954. In that year a driverless electric cart, made by Barrett Electronics Corporation, began pulling loads around a South Carolina grocery warehouse. Such machines, dubbed AGVs (Automatic Guided Vehicles), commonly navigate by following signal-emitting wires entrenched in concrete floors. In the 1980s AGVs acquired microprocessor controllers that allowed more complex behaviours than those afforded by simple electronic controls. In the 1990s a new navigation method became popular for use in warehouses: AGVs equipped with a scanning laser triangulate their position by measuring reflections from fixed retro-reflectors (at least three of which must be visible from any location).

Although industrial robots first appeared in the United States, by the 1980s companies in Japan and Europe began to vigorously enter the field. The prospect of an aging population and consequent worker shortage, as well as high labour costs, have encouraged the adoption of robot substitutes.

FULLERENES

Fullerenes are hollow carbon molecules that form either a closed cage (in which case they are known as "buckyballs") or a cylinder (carbon "nanotubes"). The first fullerene was discovered in 1985 by Sir Harold W. Kroto of the United Kingdom and by Richard E. Smalley and Robert F. Curl Jr. of the United States. Kroto, working with colleagues at the University of Sussex, Brighton, Eng., was using laboratory microwave spectroscopy techniques to analyze the spectra of carbon chains. These measurements later led to the detection, by radioastronomy, of chainlike molecules consisting of 5 to 11 carbon atoms in interstellar gas clouds and in the atmospheres of carbon-rich red giant stars. On a visit to Rice University, Houston, Texas, in 1984, Curl, an authority on microwave and infrared spectroscopy, suggested that Kroto see an ingenious laser–supersonic cluster beam apparatus developed by Smalley. The apparatus could vaporize any material into a plasma of atoms and then be used to study the resulting clusters of atoms. During the visit, Kroto realized that the technique might be used to simulate the chemical conditions in the atmosphere of carbon stars and so provide compelling evidence for his conjecture that the chains originated in stars. In a now-famous 11-day series of experiments conducted in September 1985 at Rice University by Kroto, Smalley, and Curl and their student coworkers James Heath, Yuan Liu, and Sean O'Brien, Smalley's apparatus was used to simulate the chemistry in the atmosphere of giant stars by turning the vaporization laser onto graphite in an atmosphere of helium gas. The study not only confirmed that carbon chains were produced but also showed, serendipitously, that a hitherto unknown carbon species containing 60 atoms formed spontaneously in relatively high abundance. Attempts to explain the remarkable stability of the C_{60} cluster led the scientists to the conclusion that the cluster must be a spheroidal closed cage in the form of a truncated icosahedron—a polygon with 60 vertices and 32 faces, 12 of which are pentagons and 20 hexagons—a design that resembles a soccer ball. They chose the imaginative name buckminsterfullerene for the cluster in honour of the American architect R. Buckminster Fuller, the designer-inventor of the geodesic domes whose ideas had influenced their structure conjecture. In 1996 the trio was awarded the Nobel Prize for their pioneering efforts.

The elongated cousins of buckyballs, carbon nanotubes, were identified in 1991 by Iijima Sumio of NEC Corporation's Fundamental Research Laboratory, Tsukuba Science City, Japan, while investigating material extracted from solids that grew on the tips of carbon electrodes after being discharged under C_{60} formation conditions. Iijima found that the solids consisted of tiny tubes made up of numerous concentric "graphene" cylinders, each cylinder wall consisting of a sheet of carbon atoms arranged in hexagonal rings. The cylinders usually had closed-off ends and ranged from 2 to 10 micrometres (millionths of a metre) in length and 5 to 40 nanometres (billionths of a metre) in diameter. High-resolution transmission electron microscopy later revealed that these multiwalled carbon nanotubes (MWNTs) are seamless and that the spacings between adjacent layers is about 0.34 nanometre, close to the spacing observed between sheets of graphite. The number of concentric cylinders in a given tube ranged from 3 to 50, and the ends were generally capped by fullerene domes that included pentagonal rings (necessary for closure of the tubes). It was soon shown that single-walled nanotubes (SWNTs) could be produced by this method if a cobalt-nickel catalyst was used. In 1996 a group led by Smalley produced SWNTs in high purity by laser vaporization of carbon impregnated with cobalt and nickel. These nanotubes are essentially elongated fullerenes.

The fullerenes, particularly the highly symmetrical C_{60} sphere, have a beauty and elegance that excites the imagination of scientists and nonscientists alike, as they bridge aesthetic gaps between the sciences, architecture, mathematics, engineering, and the visual arts. Prior to their discovery, only two well-defined allotropes of carbon were known—diamond (composed of a three-dimensional crystalline array of carbon atoms) and graphite (composed of stacked sheets of two-dimensional hexagonal arrays of carbon atoms). The fullerenes constitute a third form, and it is remarkable that their existence evaded discovery until almost the end of the 20th century. Their discovery has led to an entirely new understanding of the behaviour of sheet materials, and it has opened an entirely new chapter of nanoscience and nanotechnology—the "new chemistry" of complex systems at the atomic scale that exhibit advanced materials behaviour. Nanotubes in particular exhibit a wide range of novel mechanical and electronic properties. They are excellent conductors of heat and electricity, and possess an astonishing tensile strength. Such properties hold the promise of exciting applications in electronics, structural materials, and medicine. Practical applications, however, will only be realized when accurate structural control has been achieved over the synthesis of these new materials.

Glossary

allograft Graft of tissue taken from a donor of the same species as the recipient, wherein the donor and recipient are sufficiently unlike genetically.

astrolabe Compact instrument used to observe and calculate the position of celestial bodies before the invention of the sextant.

celluloid Tough, flammable thermoplastic composed essentially of cellulose nitrate and camphor.

Code of Hammurabi Well-preserved ancient code of law, created around 1790 BCE in ancient Babylon. It was enacted by Hammurabi, the sixth Babylonian king.

epigraphy Study of ancient inscriptions.

flange A rib or rim for strength, for guiding, or for attachment to another object.

hieroglyphic Written in or belonging to a system of writing mainly in pictorial characters.

homeostasis Relatively stable state of equilibrium or a tendency toward such a state between the different but interdependent elements or groups of elements of an organization, population, or group.

kilopascal 1,000 pascals; a pascal is a unit of measure in the meter-kilogram-second system equivalent to one newton per square meter.

ligature The action of binding or tying.

lingua franca Any of various languages used as common or commercial tongues among peoples of diverse speech.

Mesopotamia Region of southwest Asia between the Tigris and Euphrates rivers extending from the mountains of east Asia Minor to the Persian Gulf.

microprocessor A computer processor contained on an integrated-circuit chip.

millrace A canal in which water flows to and from a mill wheel.

paleography The study of ancient writings and inscriptions.

pathogenic Causing or capable of causing disease.

piezoelectric Of, relating to, marked by, or functioning by means of electricity or electric polarity due to pressure, especially in a crystalline substance such as quartz.

plenum Air-filled space in a structure, especially one that receives air from a blower for distribution through ventilation.

pneumatic Moved or worked by air pressure.

pontil Metal rod used for fashioning hot glass.

riparian Relating to or living or located on the bank of a river.

serotype Group of intimately related microorganisms distinguished by a common set of antigens.

Silicon Valley Southern part of the San Francisco Bay Area in Northern California, named for the region's many silicon chip manufacturers.

stylus An instrument for writing, marking, or incising.

treadle Swiveling or lever device pressed by the foot to drive a machine.

trompe l'oeil Style of painting in which objects are depicted with photographically realistic detail.

For Further Reading

Anderson, Romola, and R.C. Anderson. *A Short History of the Sailing Ship*. Mineola, NY: Dover Publications, 2003.

Bliss, Michael. *The Discovery of Insulin*. Chicago, IL: University of Chicago Press, 2007.

Brown, Terry A. *Gene Cloning and DNA Analysis: An Introduction*. Malden, MA: Blackwell Publishing, 2006.

Cardwell, Donald. *Wheels, Clocks, and Rockets: A History of Technology*. New York, NY: W. W. Norton & Co., 2001.

Carlisle, Rodney. *Scientific American Inventions and Discoveries: All the Milestones in Ingenuity from the Discovery of Fire to the Microwave Oven*. Hoboken, NJ: John Wiley & Sons, 2004.

Crowley, David, and Paul Heyer. *Communication in History: Technology, Culture, Society*. Boston, MA: Allyn & Bacon, 2006.

Crump, Thomas. *A Brief History of the Age of Steam: From the First Engine to the Boats and Railways*. Philadelphia, PA: Running Press, 2007.

Durr, Kenneth, and Lee Sullivan. *International Harvester, McCormick, Navistar: Milestones in the Company That Helped Build America*. Portland, OR: Graphic Arts Center Publishing Co., 2007.

Eisenstein, Elizabeth L. *The Printing Revolution in Early Modern Europe*. New York, NY: Cambridge University Press, 2005.

Glassner, Jean-Jacques. *The Invention of Cuneiform: Writing in Sumer*. Baltimore, MD: Johns Hopkins University Press, 2007.

Goetz, Alisa. *Up, Down, Across: Elevators, Escalators, and Moving Sidewalks*. New York, NY: Merrell Publishers, 2003.

Gunston, Bill. *The Development of Jet and Turbine Aero Engines*. Somerset, UK: Haynes Publishing, 2006.

Hayes, Walter P., and Michael C Barnes. *Dams: Impacts, Stability, and Design*. Hauppague, NY: Nova Science Publishers, 2009.

Jazar, Reza N. *Theory of Applied Robotics: Kinematics, Dynamics, and Control*. New York, NY: Springer, 2007.

Kirby, Maurice W. *The Origins of Railway Enterprise: The Stockton and Darlington Railway 1821–1863*. New York, NY: Cambridge University Press, 2002.

Marsden, Ben. *Watt's Perfect Engine: Steam and the Age of Invention*. New York, NY: Columbia University Press, 2004.

Rose, Alexander. *American Rifle: A Biography*. New York, NY: Delacorte Press, 2008.

Schwartz, Evan I. *The Last Lone Inventor: A Tale of Genius, Deceit, and the Birth of Television*. New York, NY: Harper Paperbacks, 2003.

Spenser, Jay. *The Airplane: How Ideas Gave Us Wings*. New York, NY: HarperCollins, 2008.

Stross, Randall E. *The Wizard of Menlo Park: How Thomas Alva Edison Invented the Modern World*. New York, NY: Three Rivers Press, 2008.

Swedin, Eric G., and David L. Ferro. *Computers: The Life Story of a Technology*. Baltimore, MD: Johns Hopkins University Press, 2007.

Yorke, Stan. *Windmills and Waterwheels Explained*. Newbury, England: Countryside Books, 2006.

Index

A

Abbe, Ernst, 327
Adler, Dankmar, 223
agriculture, advancements in, 15, 145, 334, 340–343, 346–350, 364–365
airplanes, invention of, 128–133
    jetliners, 17, 137–141
Alexanderson, Ernst F.W., 78
Allen, Willard, 254
alphabet, invention of
    Greek, 29–32, 33
    Latin, 32–34
Altair, 91, 94
aluminum, discovery of, 354–356
American Telephone & Telegraph Company (AT&T), 61, 75–76, 78, 83
antibiotics, discovery of, 18, 244–246
Appert, Nicolas, 346–347
Apple Inc., 91, 92
Apple II, 91, 92, 94
aqueduct, invention of, 17, 201–203
Arber, Werner, 263
arch, invention of, 17, 192–196, 197, 204
Arkwright, Richard, 158
Armat, Thomas, 71
ARPANET, 95–98
arrow, invention of, 277
artificial heart, 258–260
Aspdin, Joseph, 212
assault rifle, invention of, 291–294
atomic clock, invention of, 331–333
Audion, 75
automobile, invention of, 17, 124–128
Avery, William, 168

B

Bacon, Francis Thomas, 191
Bacon, Roger, 325
Baekeland, Leo Hendrik, 361–363
Baeyer, Adolf von, 362
Baird, John Logie, 77–78, 82
Bakelite, invention of, 361–364
ballistic missile, invention of, 298–303
Banting, Fredrick, 243
Bardeen, John, 83, 85–86, 87
Barnard, Christiaan Neethling, 256, 261
Basov, Nikolay Gennadiyevich, 181
battery, invention of electric, 15, 158–161
Bauer, Edmond, 190–191
Bayer, Karl Joseph, 356
Beadle, Clayton, 360
Beau de Rochas, Alphonse, 125, 171–172
Becquerel, Antoine-César, 187
Beeldsnijder, Francois, 327
Bell, Alexander Graham, 54, 55, 56, 57–58, 60–62, 63–64
Bell, Chichester, 63–64
Bell Laboratories, 83, 84–86, 87, 181, 183
Bennett, William, Jr., 183

375

Benz, Karl, 125, 126–127, 128, 173
Berg, Max, 213
Berliner, Emil, 54, 64, 67
Berners-Lee, Tim, 100
Bessemer, Henry, 351, 352
Bessemer converter, 351–353, 354
Best, Charles, 243
Bevan, Edward J., 360
bicycle, invention of, 17, 121–124
Billings, John and Evelyn, 254
birth control pill, invention of, 253–256
Blenkinsop, John, 118
blood transfusion, invention of, 246–250
Blundell, James, 246–247
Bohr, Niels, 304, 331
Book of the Dead, The, 27
Boulton, Matthew, 155
Bourseul, Charles, 55
bow and arrow, invention of, 273–277
Boyer, Herbert W., 263
Boyle, Robert, 153
Brahe, Tycho, 316
Bramah, Joseph, 209
Branly, Edouard, 73–74
Brattain, Walter H., 83, 85–86, 87
Braun, Wernher von, 298, 300
brick, invention of, 196–197
Briggs, Robert W., 265
British Broadcasting Corporation (BBC), 78, 82
Broers, G.H.J., 191
bronze, development of, 334–337
Bronze Age, 335–336
Brown, Robert, 326

building construction, advancements in, 17, 192–230
Burnet, Frank Macfarlane, 252
Burnham, Daniel, 224
Buseman, Adolph, 139
Bush, Vannevar, 100
Bushnell, David, 284, 286

C

Cai Lun, 37
Calley, John, 156
camera obscura, 47, 48, 49
canning, invention of, 346–348
Carbutt, John, 69
Carlisle, Anthony, 160
Carnot, Sadi, 171
Carré, Ferninand, 349
Carrel, Alexis, 257
Carrier, Willis, 229
Cartwright, Edmund, 345–346
Cassegrain, Laurent, 322
CAT scanning, 241–242
Cavendish, Henry, 161
Cayley, George, 130, 138
Cerf, Vinton, 99
CERN, 100
Chain, Ernst, 244
Chapin, Daryl, 188
Chappe, Claude and Ignace, 51
Chardonnet, Hilaire Bernigaud, comte de, 359–360
Chinese junk, 108
chloroform, 236, 237
Chrysler Building, 219
cinématographe, 71
civil engineering, advancements in, 17, 192–230

Clark, Alvan, 324
Clarke, Arthur C., 88
Clavius, Christopher, 315
climate control, 227–230
clock, invention of, 13, 317–319
cloning, invention of, 264–271
   reproductive, 266–269, 271
   therapeutic, 269–271
codex, invention of, 35, 36–37
Cohen, Stanley N., 263
Coignet, François, 212–213
Colburn, Irving, 359
Colton, Gardner, 235
Columbia Phonograph
   Company, 66–67, 81
combine harvester, invention of,
   13, 15, 364–365
communication, advancements
   in, 15–17, 21–101
communications satellite,
   invention of, 87–90
Compaq Computer
   Corporation, 93
compass, invention of, 109–112
computer, invention of personal,
   14, 90–101
Computer Science Network
   (CSNET), 98
Constant, H., 176
Contamin, Victor, 219
Cooke, Sir William Fothergill, 53
Coolidge, George, 164
copper, use of, 334, 336–337
Corbusier, Le, 220, 221
Cormack, Allan, 241–242
Cornelisz, Cornelis, 151
Cort, Henry, 340
Coulomb, Charles-Augustin
   de, 160

Cowen, James, 295
Cros, Charles, 63
Cross, Charles F., 360
crossbow, invention of, 18,
   275–276
Cullen, William, 348–349
cuneiform, invention of, 21–25, 26
Curl, Robert F., Jr., 367, 368
Curtis, Charles G., 169

D

Daguerre, Louis-Jacques-Mandé,
   49–51
daguerreotype, 49–51
Daimler, Gottlieb, 125, 127–128, 173
dams, invention of, 198–200
Darby, Abraham, 339
Davtyan, O.K., 191
Davy, Sir Humphrey, 160, 164,
   235, 354
DeBakey, Michael E., 259
Deere, John, 343
Defense Advanced Research
   Projects Agency (DARPA),
   95, 99
De Forest, Lee, 74–75
de Laval, Carl G.P., 169
della Porta, Giambattista, 47
Delvigne, Henri-Gustave, 281, 282
Denis, Jean-Baptiste, 246
Despeissis, Louis-Henri, 360
Deville, Henri Sainte-Claire, 355
Devol, George, 366
DeVries, William C., 259
D'Haenens, Irnee, 183
Dickinson, John, 40
Dickson, W.K.L., 70, 71
diode, invention of, 74

Doisy, Alan, 254
Dolly (cloned sheep), 265–266, 267, 268–269
Dornberger, Walter, 298, 300
Drebbel, Cornelis, 284
Drew, Charles Richard, 248
Dubos, René, 245
Dunlop, John Boyd, 124
Durand, Peter, 347
Dutert, C.-F.-L., 219
dynamite, invention of, 216–219

E

Eastman, George, 50, 69, 70, 363
Easton, Edward D., 66
Edison, Thomas
  and the light bulb, 164
  and motion picture, 69–70, 71
  and the phonograph, 62, 63, 64, 66–67, 68
  and the radio, 74
  and the telephone, 54, 62
Eiffel, Gustave, 219
Eiffel Tower, 219
Einstein, Albert, 180–181, 304
Electric and Musical Industries, Ltd. (EMI), 81–82
electric battery, invention of, 15, 158–161
electric generator and motor, invention of, 161–163, 185
electrolysis, discovery of, 160
electromagnetism, 53, 72, 161–162
electron tube, invention of, 76
elevator, invention of, 225–227
Elling, Aegidus, 174
Enders, John, 252
energy, advancements in, 14–15, 144–191

Engelbart, Douglas, 100
Engelberger, Joseph, 366
Ericsson, John, 120
ether, 235, 236
Evans, Oliver, 349

F

Fairbairn, William, 148
Fair Store, 224
Fantus, Bernard, 250
Faraday, Michael, 53, 72, 160, 161–162, 163
Farnsworth, Philo, 80, 81, 82
fermentation, 237–238
Fermi, Enrico, 178, 305, 310
fire, ability to control use of, 13, 14, 15, 144–145
fire drill, 144, 145
Fitch, John, 113–114
Fleming, Alexander, 244
Fleming, Sir John Ambrose, 74
Florey, Howard, 244
fluorescent light, 165–166, 167
Fourcault, Émile, 358
Fourdrinier, Henry and Sealy, 40
Fracastoro, Girolamo, 325
Francis, Thomas, Jr., 253
Franz, Anselm, 177
Freon, 350
Freyssinet, Eugène, 213
Frisch, Otto, 303–304
Fritts, Charles, 187
fuel cell, invention of, 13, 15, 190–191
Fuller, Calvin, 188
Fuller, R. Buckminster, 368
fullerenes, discovery of, 367–370
Fulton, Robert, 114–116

Fyodorov, Vladimir Grigorevich, 292

G

Galileo, 319, 322
Gallery of Machines, 219–221
Galvani, Luigi, 158–160
galvanism, 159
gasoline engine, invention of, 125–126, 127–128, 170–173
general anesthesia, development of, 235–237
General Electric Company (GE), 78, 164, 166–167, 183
generator, invention of electric, 161–162, 185
genetic engineering, invention of, 13, 15, 262–264
Gibbon, John H., Jr., 258
glass, invention of sheet and plate, 356–359
Glass, Louis, 66
glass curtain wall, 220–221, 224
Goddard, Robert H., 134
Gorrie, John, 349
Gottschalk, Felix, 66
Gould, Gordon, 181–182, 183
GPS (global positioning system), invention of, 141–143
Gramme, Zénobe Théophile, 162
gramophone, invention of, 64
graphical user interface (GUI), 92–93
graphophone, invention of, 63–64
Gray, Elisha, 56–60
Gregorian calendar, invention of, 314–317
Gregory, James, 322, 323

Gregory XIII, Pope, 314, 315
Griffith, A.A., 176
Grove, William, 190
Groves, Leslie R., 306
Guericke, Otto von, 154
Guido da Vigevano, 295
gunpowder, invention of, 278–280
guns, invention of, 13, 18, 280–283, 288–294
Gurdon, John, 265
Gutenberg, Johannes, 16, 40, 43, 44–45
Gutenberg Bible, 44–45
Guthrie, Charles, 257

H

Haber, Fritz, 190–191
Haberlandt, Ludwig, 254
Hale, George Ellery, 323
Hale Telescope, 323, 324–325
Hall, Charles M., 355–356
Hall, Chester Moor, 327
Hall, Robert N., 183
Hallay, David, 153
Hargrave, Lawrence, 138
Harington, Sir John, 209
Harrison, James, 349
Harrison, Wallace, 221
Harvey, William, 246
Heape, Walter, 254
heart, artificial, 258–260
heart, mechanical, 258, 259–260
heart-lung machine, 258–259
heart transplantation, invention of, 256–262
heating, ventilation, and air conditioning (HVAC), invention of, 227–230

Heinkel, Ernst, 175, 176
heliography, 48, 51
Henlein, Peter, 319
Hennebique, François, 213
Henry, Joseph, 53
Heron of Alexandria, 167–168
Héroult, Paul-Louis-Toussaint, 355–356
Herriot, Donald, 183
Herschel, Sir William, 323–324
Hertz, Heinrich, 73, 328
Hoagland, Hudson, 255
Holabird, William, 220, 223, 224
Holland, John P., 286–287
Home Insurance Company Building, 221–223, 224
Hooke, Robert, 321, 325
hormones, discovery of, 242
Hounsfield, Godfrey, 241–242
Humboldt, Alexander von, 160
Huygens, Christiaan, 319, 321
Hwang Woo Suk, 270–271
Hyland, L.A., 329

I

IBM Corporation, 92, 93
IBM PC, 92
Iijima Sumio, 369
Illig, Moritz Friedrich, 39
Image Dissector, 80–81
Industrial Revolution, 13, 15, 117, 146, 200, 228
industrial robots, invention of, 366–367
industry, advancements in, 334–370
Information Age, 14, 83
insulin, discovery and use of, 242–244

Intel Corporation, 91, 92, 93
Internet, invention of, 13, 14, 16–17, 95–101
iron, use of, 337–340
Iron Age, 336, 337, 338, 340

J

Jackson, Charles, 236
Jansen, Hans, 325
Jansen, Zacharias, 325
Janszoon, Laurens, 42
Jarvik, Robert K., 259–260
Javan, Ali, 183
Jenkins, Charles Francis, 77–78
Jenner, Edward, 233, 234
Jenney, William Le Baron, 221–223, 224
Jerome, St., 29, 35
jet engine, invention of, 173–177
jetliners, invention of, 17, 137–141
Jobs, Steven P., 91, 94
Johnson, Charles, 212
Johnson, Eldridge, 67
Jouffroy d'Abbans, Claude-François-Dorothée, marquis de, 113

K

Kahn, Robert, 99
Kalashnikov, Mikhail Timofeyevich, 293
Kay, John, 345
Kekwick, Alan, 249
Kelly, Mervin, 84–85
Kelly, William, 352
Kepler, Johannes, 316
Ketelaar, J.A.A., 191

keystone, 192
Kinescope, 80
Kinetograph, 70
Kinetoscope, 70–71
King, Thomas J., 265
Kling, Carl, 251
Knaus, Hermann, 253–254
Kodak camera, 50, 69
Kroto, Sir Harold W., 367–368

L

LaCour, P., 185
Ladenburg, Rudolf Walther, 181
Lake, Simon, 286, 287
Landois, Leonard, 247
Landsteiner, Karl, 247–249, 252
Langer, Carl, 190
Langerhans, Paul, 242
laser, invention of, 180–184
Lassell, William, 324
Laubeuf, Maxine, 284
LeBlanc, Maurice, 77
Leeuwenhoek, Antonie van, 326
Leiter Building, 224
Leith, Emmett, 183
Lenoir, Étienne, 125, 171, 172
Leonardo da Vinci, 13, 295
Levassor, Émile, 128
Leyden jar, 159
light bulb, invention of incandescent, 15, 163–167
Lilienthal, Otto, 138
Lilio, Luigi, 315
Lippershey, Hans, 325
Lippincott, Jesse, 64–66, 67
Lister, Joseph Jackson, 327
Liverpool and Manchester Railway, 119–121
Livingston, Robert, 115–116
Long, Crawford, 236
Lower, Richard, 246
Ludington Building, 223, 224
Lumière, Auguste and Louis, 71

M

Macdonald, Thomas H., 66
machine gun, invention of, 288–291
Macintosh, 93
Macnamara, Jean, 252
Maiman, Theodore H., 182–183
Maltese cross, 69, 70
Manhattan Building, 223, 224
Manhattan Project, 178, 305–307, 310
Marconi, Guglielmo, 72, 73, 75, 76
Marcus, Siegfried, 125, 126, 127
Marey, Étienne-Jules, 69
Marriott, Hugh, 249
Marshall, John, 254
Martin, Benjamin, 327
Martin, Pierre and Émile, 351
Maxim, Hiram Stevens, 288–290
Maxwell, James Clerk, 72–73, 328
May, Joseph, 76
Maybach, Wilhelm, 127–128, 173
McAdam, John Loudon, 211–212
McCormick, Katharine, 254
measurement, advancements in, 18, 314–333
medicine, advancements in, 17, 18, 231–271
Meerwein, Karl Friedrich, 129
Mering, Joseph von, 243
Michaux, Ernest, 121
Michaux, Pierre, 121

Micrographia, 325–326
microscope, invention of, 18, 325–327
Microsoft Corporation, 92, 93, 101
Microsoft Windows, 93
Mies van der Rohe, Ludwig, 220, 221
military technology, advancements in, 18–19, 272–313
Minié, Claude-Étienne, 281–282
Minkowski, Oskar, 243
Moleyns, Frederick de, 164
Mond, Ludwig, 190
Monier, Joseph, 213
Moore, Hiram, 15
Moore-Brabazon, John, 139
Morse, Samuel F.B., 53–54
Morse code, 53–54, 73, 74
Morton, William Thomas, 235, 236
motion pictures, invention of, 68–72
motor, invention of electric, 15, 163, 284
Mouchel, Louis, 213
movable type, invention of, 13, 16, 40, 41, 43–45
Murray, George, 52
Musschenbroek, Pieter van, 159
Muybridge, Eadweard, 68

N

Nathans, Daniel, 263
National Phonograph Company, 67
Nedelin, Mitrofan Ivanovich, 302
Nelmes, Sarah, 233
Nelson, Ted, 100
Nernst, Walther H., 190–191

Netscape, 100–101
"networked individualism," 97–98
Newcomen, Thomas, 154–155, 156
Newton, Sir Isaac, 322
Nicholson, William, 160
Nielsen, James Beaumont, 340
Niépce, Nicéphore, 48, 49, 51
Nipkow, Paul, 77
nitrous oxide, 235–236
Nobel, Alfred, 216–219
Nobel, Emil Oskar, 217
Nobel, Immanuel, 217
Nordenfelt, Torsten, 286
noria, 148
nuclear reactor, invention of, 178–180
nuclear weapons, invention of, 18, 303–313

O

Oberth, Hermann, 134
observation, advancements in, 18, 314–333
Ogino, Kyusaku, 253–254
Ohain, Hans von, 138, 175, 176
Ohl, Russell, 188
Ohm, Georg Simon, 160–161
Ohm's law, 161
Old Colony Building, 223
Olivier, René and Aimé, 121–122
Oppenheimer, J. Robert, 306–307, 311
ornithopter, 129–130, 132
Ørsted, Hans Christian, 53, 354
Ostwald, Wilhelm, 190
Otis, Elisha Graves, 225
Otlet, Paul, 100
Otto, Nikolaus August, 125, 127, 172–173, 223

P

Panhard, René, 128
Pantheon, 205–206, 207, 213
paper, invention of, 37–40
Papin, Denis, 154
papyrus plant, 28
papyrus scroll, invention of, 16, 25–29, 34
   and Egyptians, 25–27, 28
   and Greeks, 27–29
parchment, invention of, 28, 34–35
Parsons, Sir Charles Algernon, 169
Parsons, William, 3rd Earl of Rosse, 324
Pasteur, Louis, 17–18, 237–238, 347
pasteurization, invention of, 17–18, 237–239
paved roads, invention of, 17, 209–212
Pearson, Gerald, 188
penicillin, 18, 244–245
penny-farthing, 123
Perry, Stewart, 153
Perskyi, Constantin, 78
personal computer, invention of, 14, 90–101
Phipps, James, 233
phonograph, invention of, 62–68
photography, invention of, 46–51
Pierce, John Robinson, 87, 88
Pilkington, Alastair, 358
Pincus, Gregory, 255
Pi Sheng, 41
plateway, development of, 117
Pliny the Elder, 25, 341, 354
plowshare, invention of, 340–343
plumbing, invention of, 17–18, 207–209
polio vaccine, invention of, 18, 250–253
Polo, Marco, 42
polyphony, 22, 23
Poncelet, Jean-Victor, 148
Pope, Albert E., 123
Portland Cement, 212
post-and-lintel system, 193
power and energy, advancements in, 14–15, 144–191
power loom, invention of, 13, 344–346
pozzolana, 204
Prescott, Samuel C., 347
printing press, invention of, 13, 16, 40–46
Prokhorov, Aleksandr Mikhaylovich, 181

R

Raciborski, Adam, 253
radar, invention of, 328–331
radio, invention of, 14, 15, 72–76
Radio Corporation of America (RCA), 80, 82
railroad system, invention of, 17, 116–121
Ransome, Ernest, 213
Rateau, C.E.A., 169
rayon, invention of, 359–361
recombinant DNA technology, 262–264
refrigeration, invention of, 15, 229–230, 348–350
reinforced concrete, invention of, 212–214
Reis, Johann Philipp, 55
Renaissance, 13, 206, 228
Renold, Hans, 124

Reynaud, Émile, 68
rifled muzzle-loaders, invention of, 280–283
Robbins, Frederick, 252
Robert, Nicolas-Louis, 39
robots, invention of industrial, 366–367
Roche, Martin, 223
Rock, John, 255
Rocket, 120
rocket propulsion, 133–137
Roebling, John Augustus, 214–216
Roebling, Washington, 216
Roger, Émile, 127
Roman dome, invention of, 204–207
Röntgen, Wilhelm Conrad, 239–241
Roosevelt, Nicholas, 116
Root, John, 220, 224
Rose, Max, 126–127
Rosing, Boris, 79–80
Rubin, Eduard Alexander, 279
Rumsey, James, 113

S

Sabin, Albert, 253
safety bicycles, 123–124
sailing ships, invention of, 104–109
sails, use of on ships, 104, 106–109
Salk, Jonas, 253
Sanger, Margaret, 254–255
Sarnoff, David, 80, 82
Savery, Thomas, 154, 156
Schawlow, Arthur L., 181, 182, 183
Scheinman, Victor, 366
Schlep, Helmut, 175–176
Schmeisser, Hugo, 292

Schöffer, Peter, 43–45
Schulze, Johann Heinrich, 47
Schweizer, Matthias Eduard, 360
Scott, C.P., 79
Scott de Martinville, Édouard-Léon, 63
semaphore, invention of, 51–52
Shockley, William B., 83, 85, 86, 87
Shoenberg, Isaac, 82
Siemens, William and Friedrich, 351
Simms, F.R., 295
Simpson, James Young, 237
skyscraper, invention of, 219–225
Smalley, Richard E., 367, 368
smallpox vaccine, development of, 231–235
Smeaton, John, 147, 151, 212
Smith, Hamilton O., 263
Sobrero, Ascanio, 217
solar cell, invention of, 15, 187–189
space launcher, invention of, 17, 19, 133–137
spear, invention of, 272–273
Spemann, Hans, 265
Starley, James, 122, 124
Starley, John Kemp, 124
steam, use of, 14–15, 118, 146, 153–158, 167–170, 226, 228–229
steamboat, invention of, 17, 113–116
steam engine, invention of, 117, 153–158, 167, 168, 170
steam locomotive, invention of, 118
steam turbine, invention of, 13, 167–170
steel, production of, 350–354
stem cells, 269–270, 271

Stephenson, George, 117–118, 119, 120
Stephenson, Robert, 120
Stock Exchange Building (Chicago), 223
Stockton and Darlington Railway, 118–119, 120
Stoner, Eugene M., 293–294
streptomycin, 245
Stringfellow, John, 138
Sturgeon, William, 53
submarine, invention of, 284–288
Sullivan, Louis, 224
suspension bridge, invention of, 17, 214–216
Swan, Sir Joseph Wilson, 164, 359
Swinton, A.A. Campbell, 79
Symington, William, 114

T

Tainter, Charles Sumner, 63–64
tank, invention of, 294–298
telegraph, invention of, 14, 16, 51–54
telephone, invention of, 14, 16, 54–62
telescope, invention of, 18, 322–325
television, invention of, 14, 15, 76–82
Telford, Thomas, 210–211
Teller, Edward, 310–313
Tesla, Nikola, 163
Thomas, Sidney Gilchrist, 352, 353, 354
Thouvenin, Louis-Étienne de, 281, 282
Titan rockets, 301–302, 303
Townes, Charles H., 181, 182, 183
transistor, invention of, 76, 82–87

transportation, advancements/inventions in, 13, 17, 102–143
Trésaguet, Pierre-Marie-Jérôme, 210
Trevithick, Richard, 117
Tsiolkovsky, Konstantin E., 134
Turtle, 284, 285–286
Twinning, Alexander C., 349
Twyford, Thomas, 209

U

Ulam, Stanislaw M., 313
ultrasound, 240
Underwood, Michael, 250–251
Underwood, William, 347
Upatnieks, Juris, 183

V

vaccines, invention of, 18, 231–235, 250–253
Vail, Alfred, 54
vault, 193, 195, 197
vellum, invention of, 28, 34–35
velocipedes, 121–122
Victor Talking Machine Company, 67, 82
virtual communities, 96–98
Vitruvius, 147, 148, 225
Voight, Henry, 113–114
Volta, Alessandro, 52, 159–160
voltaic cell, invention of, 52
voltaic pile, 159, 160
voussoir, 192

W

Wagner, Herbert, 175–176
Wang Chen, 41
watch, invention of, 319–322

water, harnessing of power of, 14, 146–149
water turbines, 149
waterwheel, invention of, 146–149, 184
Watson, Thomas, 54, 57, 60
Watson-Watt, Sir Robert, 330
Watt, James, 114, 155, 168, 228
Weller, Thomas, 252
Wells, Horace, 235–236
Wenström, Jonas, 162
Westinghouse, 80, 163, 166–167
Wheatstone, Sir Charles, 53
wheeled chariot, invention of, 102–104
Whittle, Frank, 138, 174–176
Wickman, Ivar, 251
Wiener, Alexander, 249
Williams, Charles, Jr., 61
Wilmut, Ian, 265, 267, 268
wind, harnessing of power of, 14, 149–153, 184–187
windmill, invention of, 146, 149–153, 184–185
wind turbine, invention of, 184–187
Wöhler, Friedrich, 354–355
World Wide Web, creation of, 100
Wozniak, Stephen G., 91, 94
Wright, Wilbur and Orville, 130–133, 138
writing, invention of, 15–16, 21–25

X

X-ray imaging, invention of, 239–242
xylography, 42

Y

Yudin, Sergey Sergeyevich, 250

Z

Zédé, Gustave, 284
zoetrope, 68, 70
Zworykin, Vladimir Kosma, 80–81, 82